CUBA'S ENERGY FUTURE

CUBA'S ENERGY FUTURE

STRATEGIC APPROACHES TO COOPERATION

Jonathan Benjamin-Alvarado
Editor

BROOKINGS INSTITUTION PRESS
Washington, D.C.

THE BROOKINGS INSTITUTION
1775 Massachusetts Avenue, N.W., Washington, D.C. 20036
www.brookings.edu

Library of Congress Cataloging-in-Publication data
Cuba's energy future : a policy assessment and strategic approaches to cooperation / Jonathan Benjamin-Alvarado, editor.
p. cm.
Includes bibliographical references and index.
Summary: "Examines what Cuba must do to ensure energy sustainability and self-sufficiency in order to secure its future, including advancing relationships with its neighbors, and discusses ways the island nation might seek greater cooperation with the United States in these endeavors"—Provided by publisher.
ISBN 978-0-8157-0342-6 (pbk. : alk. paper)
1. Energy policy—Cuba. 2. United States—Foreign economic relations—Cuba. 3. Cuba—Foreign economic relations—United States. I. Benjamin-Alvarado, Jonathan. II. Brookings Institution.
HD9502.A4C92 2010
333.79097291—dc22 2010030094

9 8 7 6 5 4 3 2 1

Printed on acid-free paper

Typeset in Minion

Composition by Circle Graphics
Columbia, Maryland

Printed by R. R. Donnelley
Harrisonburg, Virginia

Contents

Preface and Acknowledgments vii

1 Evaluating the Prospects for U.S.-Cuban Energy Policy Cooperation 1
Jonathan Benjamin-Alvarado

2 Extracting Cuba's Oil and Gas: Challenges and Opportunities 21
Jorge R. Piñón and Jonathan Benjamin-Alvarado

3 The Electric Power Sector in Cuba: Ways to Increase Efficiency and Sustainability 48
Juan A. B. Belt

4 Energy Balances and the Potential for Biofuels in Cuba 80
Ronald Soligo and Amy Myers Jaffe

5 Prospects for U.S.-Cuban Energy Engagement: Findings and Recommendations 110
Jonathan Benjamin-Alvarado

Contributors 131

Index 133

Preface and Acknowledgments

Policymakers, scholars, and analysts have pondered for nearly a half century this question: what would a Cuban government look like if it were fully recognized by the United States and engaged in robust economic relations with its large northern neighbor? While we've been waiting for an answer, intervening events, such as the end of the cold war and the peaceful transfer of power in Cuba to Fidel Castro's brother, have not prompted any change in the prevailing status quo. The question remains relevant because of the sustained power and allure that this strategically important island poses in a region rife with uncertainty, and where the stakes for regional progress, peace, and development rest in part with Cuba's own development. As the island nation seeks to capitalize on its new-found oil reserves and as the United States casts about for energy sources outside the Middle East, these two countries could come together in a manner inconceivable just five years ago.

This volume is an outgrowth of several events that have probed Cuba's energy past and future in depth, beginning with the 2006 annual meeting of the Association for the Study of the Cuban Economy (ASCE), where I served as discussant for a paper delivered by former oil executive Jorge R. Piñón titled "Energy: Cuba's Achilles Heel." I sat on a panel with Piñón and Juan Belt, one of the Americas' leading electricity infrastructure analysts. After the meeting, I asked my ASCE colleagues if we couldn't continue to refine our respective analyses and present the findings at two Cuba-related conferences coming up in 2008. Both Piñon and Belt agreed enthusiastically. Ron Soligo and Amy Myers Jaffe of the James Baker Institute at Rice University were added to broaden our analysis to include the prospects of energy demand in Cuba under varying scenarios.

At the second of these meetings, we met with Ambassador Vicki Huddleston (former head of the U.S. Interests Section in Havana and now with the Defense Department), then leading the Brookings project on U.S. Policy toward a Cuba in Transition with Ambassador Carlos Pascual (former vice president and director for Foreign Policy at Brookings and now U.S. ambassador to Mexico), to learn how our own efforts might contribute to the Brookings work.

Under the auspices of this Brookings Institution project, a group of distinguished academics, opinion leaders, and international diplomats had been brought together to seek pathways to a strong and effective U.S. policy toward Cuba. The group's final report concluded that the United States should adopt a policy of critical and constructive engagement, phased in unilaterally. After learning more of our work in the energy sphere, Huddleston, Pascual, and their colleagues working with Brookings readily agreed to publish our findings.

This volume, therefore, is an extension of the Brookings project's analysis by looking concretely at the strategic and material challenges posed by Cuba as a potentially ascendant, albeit mid-level, oil state, and how it might serve as a partner in strategic regional objectives, in terms of both national and energy security interests of the United States.

No book can be launched without committed contributors, and I would first like to thank Jorge Piñón, Juan A. Belt, Ronald Soligo, and Amy Myers Jaffe. They represent an august body of experts on the subject of Cuban energy issues. Their collective efforts have made this volume an important contribution to the specifics of Cuban energy development and to the broader discussion of the possibility of a sober and objective dialogue and cooperation with the Cuban regime.

Our colleagues at the Brookings Institution—Carlos Pascual, Vicki Huddleston, Ted Piccone, senior fellow and deputy director for Foreign Policy, and especially, Dóra Beszterczey, former research assistant for the U.S. Policy toward a Cuba in Transition project, have been instrumental in coordinating our efforts and guiding us through the editorial process.

If the sharing of time and information by former and present officials of Cuba's energy-related bureaucracies is any indication of what the future of U.S.-Cuban relations might be like, our experience portends an honest, open, and mutually beneficial relationship. Time and again, we have been pleasantly surprised by the forthrightness of Cuban officials to discuss all facets of energy questions. It must be noted that without their cooperation and information sharing the depth of this analysis would be all but unreachable. In particular, I would like to thank Eloy Leon Gomez, Manuel Marrero Faz, Rafael Tenreyro Perez, Raul Perez de Prado, Juan Fleites Melo, Alfredo Curbelo Alonso, and Fidel Castro Diaz-Balart for support in helping us to understand the totality of the Cuban energy picture.

I have traveled widely throughout Latin America to collect data and information on Cuba's energy profile and have been in direct contact with Cuban officials representing all of the major actors in energy-related pursuits, including CITMA, the Cuban Ministry of Science, Technology, and the Environment; GEPROP, the Center for the Management of Prioritized Projects and Programs; MINBAS, the Cuban Ministry of Basic Industry; Unión Eléctrica, Cuba's state electrical utility firm; Unión Cubapetróleo (Cupet), the Cuban national oil company; and the Cuban Council of State, the supreme ruling body in the Cuban government.

Here in the United States I owe a special debt of gratitude to Jorge Perez-Lopez, Damian Fernandez, and Mauricio Font for giving us the public forums that allowed us to distill our thinking and arrive at conclusions. There can be no doubt that the comments, suggestions, and criticism offered by our external readers, Charles Ebinger, Jorge Perez-Lopez, and Jorge Sanguinetty, made this book much more cogent and concise. I would also like to extend my gratitude to Casey Logan and John Preisinger, as well as Katherine Scott on the Brookings team, whose copyediting skills helped to make the chapters much more readable. I would also like to thank the Brookings Institution Press staff for making what at times is an arduous process into a highly professional and satisfying experience. From editing, to book cover design, to marketing, they deserve much credit for the presentation of the final product. That said, any errors, sins, and omissions are mine alone.

Closer to home, thanks to my colleagues at the University of Nebraska–Omaha for their support, especially Loree Bykerk and Lourdes Gouveia. Finally, to Beth Ann and Isabel Belén, my wife and daughter, to whom I am deeply indebted for allowing me the luxury and the latitude to pick up and leave often in pursuit of this Cuban grail—*muchisimas gracias.*

Jonathan Benjamin-Alvarado

CUBA'S
ENERGY FUTURE

one

Evaluating the Prospects for U.S.-Cuban Energy Policy Cooperation

Jonathan Benjamin-Alvarado

The last thing American energy companies want is to be trapped on the sidelines . . . while European, Canadian and Latin American rivals are free to develop new oil resources at the doorstep of the United States.

Simon Romero, "Spanish Seek Oil Off Cuba, As Americans Watch Silently," *New York Times,* July 7, 2004

As history shows, national security and economic prosperity are inseparable. The simplest answer—undoubtedly still complicated—is finding and drilling more oil from domestic sources, using less oil overall and importing far less than we do today. This requires a national energy strategy that has never existed, one that shifts U.S. consumption from fossil fuels like oil and coal toward carbon-freer solutions like nuclear, wind and solar.

"Security Case for a National Energy Plan," *Dallas Morning News,* editorial, June 19, 2009

These two quotations, dated nearly five years apart, distill the essence of why the United States should be looking carefully to the development of energy resources in Cuba. For the past fifty years, U.S. policy toward Cuba has relied on the application of cold war measures—economic sanctions, technology denial, and political isolation—in an effort to push Cuba over the tipping point of regime collapse and toward the peace and prosperity that would follow from embracing democracy.

This policy, which endures in part to maintain the notion that such measures will foster political change on the island and after nearly half a century is almost quaint, has caused the United States to overlook many of the tectonic shifts that have taken place in Cuba.[1] Such a singular focus has in some respects blinded U.S. policymakers to broader strategic changes in the region involving issues that can hardly be understood, much less resolved, without a Cuban presence. These policy areas include immigration, trafficking in human beings and narcotics, economic development, and now, energy. These remarks should not be taken as a suggestion to disregard the international community's long-standing demand for the Cuban government to expand personal liberties, support the rule of law, and extend human rights to all inhabitants of the island, but these demands should be balanced against equally important and perhaps more pressing economic and environmental concerns.[2]

It is relevant to U.S. energy security and geostrategic interests that 77 percent of proven oil reserves globally are held by national oil companies (NOCs) and that 11 percent of proven oil reserves are held by NOCs with equity access, meaning that these firms retain the contractual rights for exploration, extraction, and production of oil drawn from those reserves. Four of the five largest oil exporters to the United States—Saudi Arabia, Mexico, Venezuela, and Nigeria—are NOCs. There is growing concern about the extent to which imports from those countries are assured, given the potential for political conflict, economic instability, and social upheaval in any or all of those states. This means that only 11 percent of proven oil reserves not already held by NOCs are presently open to international oil companies (IOCs), many of which are based in the United States.[3] This political and economic reality heightens the potential importance of U.S. cooperation with Cuba on the issue of energy development.

At present Cuba possesses an estimated 4.6 million barrels of oil and 9.3 TFC (total final consumption) of natural gas in North Cuba Basin.[4] This is approximately half of the estimated 10.4 billion barrels of recoverable crude oil in the Alaska Natural Wildlife Reserve. If viewed in strictly instrumental terms—namely, increasing the pool of potential imports to the U.S. market by accessing Cuban oil and ethanol holdings—Cuba's oil represents little in the way of absolute material gain to the U.S. energy supply. But the possibility of energy cooperation between the United States and Cuba offers significant relative gains connected to the potential for developing production-sharing agreements, promoting the transfer of state-of-the-art technology and foreign direct investment, and increasing opportunities for the development of joint-venture partnerships, and scientific-technical exchanges.

The relative gains from increased commercial and technical cooperation obviously increases Cuba's domestic energy capacity, but it also possesses the potential of enhancing the United States' energy security by deepening its links in the region. The future vitality of energy security requires access to energy export markets but also the diffusion and dispersion of technology, innovation, research and development of enhanced productive capacities, alternative energy technologies, and the effective management of resources across the region. The economist Jeremy Rifkin argues that "distributive energy markets," marked by highly collaborative efforts to integrate diverse energy resources based in various proportions everywhere, will come to replace the prevailing model of the highly concentrated, conventional energy elites—coal, oil, natural gas, uranium—which are now found in limited geographical regions and are finite.[5]

The development of Cuba as an energy partner will not solve America's energy problems. But the potential for improving energy relations and deepening collaborative modalities with other regional partners is enhanced by pursuing energy cooperation with Cuba for two principal reasons.

1. Cuba's increasing leadership role in the Caribbean region and Central America might be used by the United States to promote collectively beneficial efforts to develop a broad range of alternative energy technologies in the Americas. A Cuba-America partnership might also serve as a confidence builder in assuaging the misgivings on the part of regional partners regarding American domination.
2. Cuba's significant human capital resources in the scientific and technological arena have been grossly underused. Cuba possesses the highest ratio of engineers and Ph.D.s to the general population of any country in Latin America, and this can been viewed as a key asset in the challenge of maintaining energy infrastructure across the region. Both Mexico and Venezuela face significant costs in maintaining their sizable energy production, refining, and storage capabilities. The integrity of these two national energy systems is of paramount interest to U.S. energy security concerns because of the potential harm to the economy that would occur if either state were unable to deliver its exports to the American market.

In this light, the impetus for normalization of relations writ large between the United States and Cuba is not oil per se, but enhanced energy cooperation, which could pave the way for technical and commercial exchanges that, given the evolving nature of energy resources and energy security, could provide an opening of collaborative efforts that could have mutually beneficial effects.

What has the failure to engage Cuba cost the United States in these geostrategic terms? Very little, one could argue. Strategically, Cuba has been a stable entity in the region. Politically, too, it has been a mostly static environment: with the embargo in place, policymakers and elected officials have been able to predict reactions to policy initiatives with relative certainty. U.S. business interests in Cuba since the early 1960s have been negligible, with the exception of a recent increase in humanitarian agricultural and medical sales. But a more central issue is this: In light of growing concerns regarding energy supplies in the United States and demands for domestic and regional exploration to meet American consumption, what is the cost to the United States of maintaining a status quo relationship with Cuba? In economic terms, the cost of the failure to engage Cuba has been considerable.

In its 2008 report, *Rethinking U.S.-Latin American Relations,* the Partnership for the Americas Commission, convened by the Brookings Institution, suggested that the basis for effective partnership between the United States and its Latin American and Caribbean partners is shared common interests. The report states, "Cuba has long been a subject of intense interest in U.S. foreign policy and a stumbling block for U.S. relations with other countries in the hemisphere."[6] Specifically, the report pinpoints two key challenges facing the region that are directly relevant to the subject of this book: securing sustainable energy supplies and expanding economic development opportunities. The April 2009 report of the Brookings project on U.S. Policy Toward a Cuba in Transition identified both medium- and long-term initiatives related to energy that directly fulfilled an element of the policy objectives recommended in their report.[7] In order to specifically promote what the report termed "a constructive working relationship with the Cuban government to build confidence and trust in order to resolve disputes, with the long-term objective of fostering a better relationship that serves U.S. interests and values," it recommended a medium-term initiative that "allows licenses for U.S. companies to participate in the development of Cuban offshore oil, gas, and renewable energy resources." The report also recommended that a long-term initiative be undertaken to "provide general licenses for the exportation of additional categories of goods and services that enhance the environment, conserve energy, and provide improved quality of life."[8]

Because of recent developments in Cuba and the growing investments being made there made by regional partners, in particular Venezuela and Brazil, the importance of Cuba's energy development objectives becomes decidedly more pronounced, in terms of both Cuba's national development priorities and the United States' energy and geostrategic priorities.[9]

One of the recommendations made in *Rethinking U.S.-Latin American Relations* is especially relevant: developing sustainable energy resources. The report recommends that the United States, in partnership with other governments in the hemisphere, establish a "Renewable Energy Laboratory of the Americas" that would promote hemispheric cooperation on developing solar, wind, and cellulosic-biomass technologies; intensify hemispheric cooperation in the peaceful use of nuclear energy; and promote regulatory regimes that are open to private energy investment and trade in energy technology and services.[10]

In a special section on U.S.-Cuban relations, *Rethinking U.S.-Latin American Relations* makes two other recommendations: "Promote knowledge and reconciliation by permitting the federal funding of cultural, academic, and sports exchanges; and encourage enhanced official contact and cooperation between U.S. and Cuban diplomats and governments."[11] The authors go on to articulate a set of steps or best practices that would serve to foster such a partnership and, more important, provide a set of measures open and flexible enough to account for the complexity and specificity of issues that surround energy development. In closing with a special section on Cuba, the report puts the spotlight on the centrality of the island nation and the effective management and potential leadership that it may offer in the effort to deal with these issues. While expanding the ambit of U.S. geostrategic interests in the region, it is critical that the discussion include the role Cuban energy development will have on the assessment and pursuit of those interests.

Cuba faces daunting policy challenges in the twenty-first century. Chief among them is the task of providing reliable sources of energy for economic development and revitalization in the post–cold war milieu. In light of the discovery of offshore oil and gas reserves, what policy trajectories and alternatives will increase the probability of energy self-sufficiency and sustainability in Cuba in the short and long term?

Perhaps at the time when Cuba diversifies its energy suppliers and develops its offshore resources it will have the economic independence necessary for political and economic evolution. As with many policy issues, Cuban energy policy may or may not conform to objectives that will lead to the successful implementation of the country's energy development objectives. The Cuban energy problem—that it is highly dependent on energy resources for its economic livelihood—is grounded in well-informed assessments, captured by the technical analyses of production capabilities, transmission and distribution challenges, and growing energy demands. This highly focused body of literature has identified significant shortcomings—high levels of dependency on imported oil, a crumbling energy production capability, and a fragile energy infrastructure—in the analysis of energy policy development and sustainability and in part acknowledges competing approaches toward the resolution of energy problems on the island.[12] But these analyses remain acutely attentive to the following elements of the Cuban reality: Cuba has learned from past experiences and is very much aware of political and economic risks related to imported oil. The collapse of the Soviet Union and the 2003 oil strike in Venezuela taught Cuba two very expensive lessons. President Raúl Castro understands the risks associated with single-source oil dependency; his visits in 2009 to Brazil, Russia, and Angola underscore his concerns. An emerging energy relationship with Brazil would provide a balance to Cuba's current dependency; other energy relationships could bring with them the possibility of corrupt and unsavory business practices.

Since the 1970s, energy development schemes in Cuba have included dedicated attempts to link such development to the material well-being of Cuban society overall. Thus, energy policy has been infused with policy initiatives in higher education, science and technological advancement, and increasing domestic human capital resources.[13] The centrality of this linkage was heightened by the significant loss of foreign assistance with the dissolution of the Soviet bloc. One could argue that this event inadvertently made Cuba independent for the first time in its history. The result of this forced independence has been the development of policy initiatives that rely on both imaginative and instrumental steps to meet the most pressing public policy needs. This includes the development of creative initiatives to enhance energy conservation, sustainability, and efficiency in organic agriculture; ecotourism; the development of energy cogeneration capabilities; and increasing offshore and onshore oil and gas production. These measures are linked to enhancing the political and economic status of the state in relation to its ability to meet the immediate basic needs of Cuban society while simultaneously pursuing economic initiatives that link Cuba to the region over the long term. Such analyses reveal how these gaps present opportunities for the United States to ameliorate bilateral relations and opportunities for the Cuban regime to grow its regional presence. However, few of these works link these evaluations to broader strategic imperatives as they apply to Cuba, the United States, and the region. Does Cuban oil hold the key to improving U.S.-Cuba relations and to facilitating rapprochement? The need to provide an answer to that question is the central premise of this book, and doing so is its objective.

Using a set of focused case study analyses of various elements of Cuba's energy sector, we explore the existing base of petroleum resources, in terms of Cuba's productive capacity and the Cuban policymakers' response to the sea change that is the discovery of offshore oil reserves.

We focus particularly on understanding the relationship of energy security, sustainability, and regime transition and continuity to energy policy and regime stability, and how they relate to a reconsideration of U.S. geostrategic interests. Cuba's ability to successfully manage its energy concerns will play a large role in the future development on the island. Speculation has been widespread as to the course the Cuban regime under Raúl Castro might pursue, and it might deviate from the policy priorities of the fifteen years from 1992 to 2007 under the leadership of Fidel Castro. The successful implementation of energy development policy in Cuba in this new policy environment can be seen as a bulwark against the political instability, crippling poverty, and economic stagnation that plague many developing states.

Regardless of how the regime evolves, the United States will be obliged to respond to these developments because of its own broader domestic commercial and energy-security interests. The success or failure of Cuba's future energy policy potentially has direct implications for policymakers in the United States because of proximity, history, and the continuing strategic importance to the United States of a stable Cuban regime and of energy security in the Caribbean region. The last factor is of course largely shaped by Cuba's strong relationship with Venezuela and its growing role in the Petrocaribe oil consortium. The contributors to this volume have conducted numerous interviews with economic, energy, and planning officials and experts in both Cuba and the United States, and have undertaken an ongoing public dissemination of the information they have gathered from this line of inquiry. As a result, the material brought together in this book can inform the policymaking and academic communities concerning the evolution and changes in energy development policy in Cuba that are directly relevant to the United States. We posited eight research questions, which fall into two groupings. The first four questions are about placing Cuban energy policy in proper context:

1. What are the best conditions for creating strong, responsive, and sustainable energy development policies in Cuba?
2. What needs must be met for Cuba to achieve a sustainable energy development policy?

3. In the post–cold war era, is a successful energy development policy in Cuba specifically tied to a particular form of government or economic development approach?
4. What explains the uneven performance (success or failure) in Cuba's energy development policy from 1992 to the present? How can these lessons be applied to the ongoing effort to develop Cuban energy resources?

The second four questions are about the U.S. foreign policy and national security dimensions of this inquiry:

1. What are the elements of a strategic energy policy? To what extent does the development of Cuban energy resources play a role in that policy? Is it positive or negative?
2. How does Cuba play into U.S. strategic energy policy, and will commercial relations (such as the trade embargo) need to be revised to achieve geostrategic and energy-security objectives?
3. How are the risks of balancing the twin objectives of U.S. energy security and satisfying U.S. energy needs best calculated, controlled, and accepted in the near and long term?
4. What is the best "timing" of U.S. policy implementation to advance the nation's strategic energy interests? Is this best done now or at a later juncture (such as after the passing of Fidel Castro or during some subsequent transfer of leadership)?

Both sets of questions are intended to help distill understanding of U.S. strategic energy policy under shifting political and economic environmental conditions in Cuba and the implications of U.S. energy policy for U.S. foreign policy in the near and long term. Because both energy policy and other policy areas can be considered "works in progress," an understanding of possible outcomes is important for future policy considerations and changes in the policymaking milieu.

Until now, little if any useful analysis has existed to aid policymakers in the design and implementation of a constructive engagement with Cuba regarding energy development. This book seeks to fill that hole by identifying, defining, and discussing the conditions under which such an engagement might occur. We will begin by setting the context of the analysis via a broad exploration of the geostrategic environment and assessing especially relevant Caribbean-based opportunities for and obstacles to energy security. This assessment includes an overview of the ongoing evolution of interstate relations in the region and the development of new modalities of international engagement by the Cuban regime under the leadership of President Raúl Castro. We also seek to identify developing opportunities for engagement and potential cleavages that may hinder the development of trade and cooperation in energy resource development.

The Geostrategic Environment of U.S. Energy Security

Those involved in managing the security interests of the United States need to understand the geostrategic implications of interstate relations in the region in terms of energy security, and the extent to which they affect cooperation between the United States and Cuba. This includes an assessment of the medium- to long-term evolution of energy cooperation between Cuba and Venezuela; of the broader relations between states aligned with the Bolivarian Alternative for the Americas (Alternativa Bolivariana para las Americas, or ALBA) and Petrocaribe consortiums; and of the growing influence of China in the region.[14] Also discussed in this volume is the extent to which the diversification and dispersion of energy resources in Cuba might be a buffer against disruptions in U.S. energy production and distribution that could result from natural disasters or market disruptions.

Before analyzing U.S. energy security in a geostrategic context, it is necessary to define "energy security" and "strategic energy policy." Energy security is the capacity to avoid disruptions caused by natural, accidental, or intentional events affecting energy and utility supply and distribution systems. Energy security is said to prevail when fuel, power production and distribution systems, and end-user devices possess the five so-called "S" characteristics, as outlined by Drexel Kleber, the director of the Strategic Operations Power Surety Task Force, in the Office of the Secretary of Defense:[15]

—*Surety.* Access to energy and fuel sources is assured.

—*Survivability.* Energy and fuel sources are resilient and durable in the face of potential damage.

—*Supply.* There is an identified available source of energy—traditional fossil fuels, alternative energy (nuclear, clean coal, biomass, landfill gas, municipal solid waste, hydrogen), or renewable energy (hydropower, geothermal pressure, wind, tidal. and solar).

—*Sufficiency.* There is an adequate quantity of power and fuel from a variety of sources.

—*Sustainability.* Operating practices can be perpetuated by limiting demand, reducing waste, and effectively exploiting alternative energy and renewable resources to the fullest extent possible.

The five "S" energy security and conservation objectives, though initially intended as a guide for the U.S. Department of Defense, have a much broader applicability; not least, they serve as value parameters for energy policy decisionmaking. As Kleber has noted, "Expenditures on energy conservation measures are viewed as 'investments' with long-term rewards and dividends which are paid in commodities beyond money—national security, soldiers' lives, improved manpower utilization, military to civilian transfers, and increased foreign policy options for elected officials, to name a few."[16]

What, then, would an ideal strategic energy policy look like for the United States—or any other country, for that matter? Mahmoud Amin El-Gamal and Amy Myers Jaffe have set out a detailed analysis of the objectives of a strategic energy policy, including the following:

1. To assure that markets operate efficiently so as to develop the infrastructure necessary to meet growing energy demand
2. To ensure the well-being of the human habitat and ecosystem
3. To ensure that mechanisms are in place for preventing and, if necessary, managing disruptions to energy supply.[17]

Articulating these objectives doesn't mean that fulfilling them is simple for policymakers for the following reasons. First, there are no overnight solutions to the energy supply and infrastructure bottlenecks facing the global markets. The trade-offs between energy-security considerations and national (non-energy) goals across the board must be continuously reviewed. States must adopt an integrated energy policy balancing foreign policy, trade policy, and national security imperatives. In this way, strategic energy policy has the ability to play a significant role in diplomatic discourse, especially where bilateral relations with major oil producers are concerned. For El-Gamal and Jaffe this is a critical consideration, for three principal reasons:

1. U.S. energy independence is not attainable.
2. The policy instruments available to deal with energy supply disruptions are increasingly inadequate.
3. The United States needs to articulate a new vision for optimal management of international energy interdependence.[18]

Thus, the questions and issues surrounding energy security become existential in a manner that has hardly been discussed heretofore, but clearly resonates in the face of ongoing changes in access to secure energy sources, persistent energy dependency, and the seemingly insatiable demand for petroleum products to fuel the American way of life. These concerns immediately raise three important questions relevant to our discussion of possible engagement with Cuba in the energy sphere:

1. How will the ongoing development and evolution of Unión Cubapetróleo S.A. (Cupet), Cuba's state oil company, limit or obstruct U.S. efforts to meet its strategic objectives?
2. What role can international oil companies play in the short- and long-term development of energy resources and infrastructure in Cuba?
3. How will the specter of competition with Brazil, Russia, China, and India over scarce petroleum resources affect U.S. energy-security policy, especially in light of the recent energy-development agreements between Brazil and Cuba, and Russia and Cuba, and the Chinese incursion into Latin American energy markets?

These questions deserve consideration, particularly in light of the growing presence of these external actors in Latin American energy markets. How might they increase competition and cooperation over scarce energy resources?

In assessing the development of Cupet and its impact on U.S. geostrategic imperatives, it is essential to evaluate how the United States might promote its interest in a global and regional energy market shaped and influenced by the activities of national oil companies, especially their influence on developments in Cuba. Including Mexico's Petróleos Mexicanos S.A. (Pemex) and Venezuela's state oil firm, Petróleos de Venezuela S.A. (PDVSA)—both of them NOCs—in this evaluation is critically important for ensuring an acceptable strategic context to U.S. interests.[19] The objective of this highly path-dependent development is the transformation of Cupet into a stable NOC that exhibits high technical competency culminating in upstream oilfield production and downstream refining and marketing capabilities. It is path-dependent because the set of decisions undertaken to achieve the objective (energy self-sufficiency) is limited by the decisions made in the past by Cuban policymakers, even though past circumstances may no longer be relevant. Prior to 2005, the energy policy objective was clearly centered on the revitalization of existing energy infrastructure and the expansion of domestic production, as limited as that may have been. Now there is a big change in Cuba's circumstances: the growing importance of tapping the offshore reserves.

An NOC, to be successful, must balance national social and political objectives with commercial objectives. Consequently, U.S. strategic policy must balance the promotion of broader U.S. interests with those of the NOC if there is to be cooperation.[20] In light of the recent resurgence of oil nationalism, future cooperation depends largely on the extent to which observers can identify and articulate the common energy-policy interests of NOCs and the United States. In Venezuela, high oil prices have encouraged the Chávez government to undertake bold social policy initiatives.[21] Some suggest these decisions have come at the expense of critical energy infrastructure needs, thereby increasing the likelihood of energy supply disruptions in the future. Because the United States relies on Venezuela for nearly 1 million barrels of oil daily, the policy decision to prioritize social spending over energy infrastructure revitalization by the Chávez regime could have a significant impact in the United States, if it were to result in diminished capacity in Venezuela to produce and export oil to the United States.[22] In Mexico, state control of the NOC Pemex has had the "stultifying impact" of prolonged bureaucratic stagnation, resulting in a decline in production and insufficient funding for reinvestment in new exploration and production. This is highly problematic for Mexico because the government derives 40 percent of its revenue from Pemex.[23] It also has raised concerns about the possibility of energy supply disruptions for the United States. In fact, in the first quarter of 2010 Mexico's oil exports to the United States fell by over 8 percent, as compared to 2009.[24]

Concerns over the ability of major oil-producing countries and their NOCs to meet future global demand is compounded by insufficient levels of reinvestment and the looming specter of interstate instability. But it is becoming abundantly clear that Venezuela's growing investment in Cuba's energy infrastructure creates the basis for a longer-term relationship that will enable Cuba to expand its productive, storage, and refining capacity, as it simultaneously strengthens the Venezuelan position in the region as a supplier of both crude and refined petroleum products for its Petrocaribe and ALBA partners.

There is also growing consternation that NOCs may be "used as instruments of state policy inimical to U.S. national interests."[25] In particular, China's growing presence in Latin America is being interpreted as a sign of intensifying competition over energy resources. Flynt Leverett and Jeffrey Bader suggest that this competition could easily be the cause of international conflict in the coming years, as energy demands place a rising premium on the ability of China—already the world's third-largest crude oil importer, after the United States and Japan—to access oil and gas resources.[26]

But Leverett and Bader also warn against an overdeterministic view of an inevitable "clash of the titans" over energy resources. Instead they argue that the impact of China on U.S. energy-security interests is a largely unexplored arena. Furthermore, one could view the Chinese search for access to oil, leading it to engage in exploration and production-sharing agreements in remote and difficult locales such as Sudan and the tar sands of Canada, as expanding the global supply of petroleum.

Developing an Analytical Framework of Cuban Energy

Our analysis in this book is based on a number of assumptions.

Assumption 1: Cuba's Energy Potential

First, we accept the U.S. Geological Survey's estimate of Cuban energy potential, as presented in its analysis of oil reserves in the Exclusive Economic Zone (EEZ), located off the north coast of Cuba. These reserves are estimated to hold 4.6 billion barrels of oil and 9.8 trillion cubic feet of liquid natural gas.[27] In assessing Cuban energy capabilities we offer a sectoral data analysis of energy production capabilities (actual and potential), including the following: upstream oil, upstream and midstream natural gas, petroleum supply and demand balance, oil marketing and convenience retailing, petrochemicals, electric power, sugarcane ethanol, and alternative energy resource potentials.

Assumption 2: Two Alternate Scenarios for Cuba

Second, we analyze the power sector in the context of two scenarios, the Business as Usual/Muddling Through Scenario and the Full Marketization Scenario. The Business as Usual/Muddling Through Scenario assumes that the essential conditions under which the Cuban economy presently functions will change little over the next decade. Under this scenario,

—Cuba will continue to receive subsidized petroleum imports from Venezuela (approximately 50 percent of the total national demand).

—The Cuban economy will rely highly on tourism for the generation of hard currency reserves.

—Cuba will continue to pursue foreign direct investment for the replacement and development of critical infrastructure but with the highly restrictive joint-venture terms under which it presently operates.

The Full Marketization Scenario assumes that Cuba will open its energy sector completely to global markets. Under this scenario, Cuba policymakers will:

—Allow foreign enterprises (international oil companies and their subsidiaries) to purchase majority shares of ownership in joint-venture projects.

—"Dollarize" its energy sector sufficiently to allow international oil and energy enterprises to buy and sell products and services inside of the Cuban economy.

—Create conditions and terms of trade that are conducive to the successful resolution of disputes over contracts such as over the rate of return on investments, penalties for early termination, and so forth.

Such steps do not imply that the Cuban economy will become completely capitalist, as Cuba may retain special prerogatives relevant to the designation of oil and gas reserves as elements of its "national patrimony." International oil companies are not averse to these types of nationalist prerogatives, which are typical of oil and gas contracts internationally.

In our analysis of the Cuban electric power sector we look at the following issues:

—Comparison with selected countries in Latin America

—Energy trends in Cuba

—Energy flows within the sector

—Financial and economic aspects of the sector

—Sector reform during a transition

—Relevant conclusions including foreign investment demands and technology transfer requirements

In chapter 2, Jorge Piñón presents an analysis of the oil and gas production resources and the capacity of the Cuban regime to successfully exploit those resources. In his power-sector analysis in chapter 3, Juan Belt employs the MARKAL/TIMES energy systems model, which generates estimates for industrial, commercial, residential, and transportation demands for energy services over the next several decades. Belt then determines whether the sources of energy will be domestic or imported by analyzing the available technologies that transform primary energy into final energy that is consumed by end-users. In chapter 4, Ron Soligo and Amy Myers Jaffe provide an energy-balance analysis for Cuba under various scenarios, including present trade and production rates, the development of offshore reserves, the development of ethanol production capabilities, and the partial and complete loss of Venezuelan oil imports. They also analyze the potential for Cuban ethanol development, including estimates of the productive capability of ethanol production, land-use and investment requirements, the potential for ethanol to offset the demand for transportation fuels, and the introduction of flex-fuel and hybrid vehicles to public and private fleets.

Assumption 3: Cuba's Projected Demand Curves

The third common assumption is based on the calculation of production and demand curves taken at three points; one in the middle term, 2015, and two in the long term, 2020 and 2025. Additionally, it estimates that the per capita GDP growth rates will be about 2, 3, and 5 percent, respectively, by means of scenarios for each year. All scenarios avoid making assumptions regarding the form or structure on the Cuban regime other than to consider the economic modalities that might accompany a general set of governing structures not limited to but including maintenance of the status quo, and perhaps a transition to a new set of leaders not including Fidel or Raúl Castro.

Applying the Analytical Framework

Using these three assumptions we divide our substantive analysis of Cuba's energy sector into four sections:

1. Cuban oil and gas characteristics
2. Cuban electric power sector requirements
3. The Cuban energy balance
4. The development of Cuban ethanol resources

In chapter 5, we summarize the findings and outline a set of recommendations for U.S. policymakers regarding the advancement of energy development cooperation, in terms of both geostrategic and economic interests.

Recommendations

Chapter 5 includes a discussion of the conditions under which the promotion of cooperative engagement between the United States and Cuba may occur, and provides an assessment of existing and perhaps new possibilities for energy cooperation in production-sharing agreements, energy resource development, technology transfer, and other mutually beneficial outcomes for the United States and Cuba in the energy sector. Following on the discussion of promoting cooperative engagement between the United States, this section will consider a set of relevant recommendations in the following thematic arenas:

Develop Confidence-Building Measures and Engagement

Despite the standoff of the past fifty years, the installation of new administrations in both Cuba and the United States creates an opportunity for the consideration of new modes of engagement that are initially symbolic and highly instrumental in nature, such as in agriculture and medical sales. Energy and infrastructure cooperation may be further areas for this type of engagement.

Create Opportunities for Leveraging Cuba's Human Capital Resources

Cuba's highly trained cadres of engineers and technicians are largely underemployed. Engaging Cuba in the areas of energy and infrastructure development may provide opportunities to employ these people and also possibly to leverage their considerable skills and abilities for cooperative projects across the region.

Transfer Energy Technology

The potential of Cuba's offshore oil reserves may only be accessible when Cuba and its partners are able to employ first-generation American deepwater exploration technology. At present, U.S. export controls limit access to this technology. Under conditions favoring resource development and production-sharing scenarios, the United States may begin to roll back these export control restrictions.

Transfer Project Management Capacity

One of the most critical findings from the analysis of Cuba's effort to develop a nuclear energy capability was the absence or notable lack of project management capacity during the design, implementation, and construction of the nuclear reactor site at Juragua.[28] Subsequent discussions with senior Cuban government officials have revealed that the development of this capacity is a high priority for Cuba as it considers the challenges it faces for future infrastructure and large construction projects. This is an area in which U.S. firms can and should play a vital role as a model and partner for Cuba. Cubans have openly expressed the desire to work side by side with American partners in this critical area of development.

Encourage Energy-Sector Trade and Cooperation

The United States and Cuba can and should cooperate in numerous areas, such as exploration, upstream production, downstream processing and distribution, transportation, and auxiliary services.

Encourage Investment and Development

There has been no lack of interest on the part of American international oil firms in developing a Cuban market for joint-venture projects and technology transfer and production-sharing agreements in the energy sector. The prevailing Cuban model of joint-venture investment and cooperation has proved to be attractive internationally, and Cuba offers American firms numerous opportunities of this type. There will have to be significant changes to the Cuban embargo before this type of engagement can occur, but recent history shows that Cuba possesses the potential to be a strong regional trade partner in the area of energy and infrastructure development. The numerous joint-venture projects presently under way in energy development and infrastructure (oil refineries, pipelines, and port facilities) between Cuba and a growing list of foreign partners is a positive indicator of that potential.

Diversify Regional Energy Resources

Creative partnerships in terms of refining, storage, and engineering services will allow the regional partners to diversify their respective portfolios, in addition to dispersing resources across the region to take advantage of location, and perhaps mitigate the potential of market disruptions owing to weather and other natural disasters.

Establish a Cuban Energy Distribution Center

A long-term prospect for Cuba may the development of energy-related resources that will be strategically positioned to serve the region's needs for oil refining and storage, oil and gas production (exploration and infrastructure), and auxiliary services. Such a distribution and services center could be a boon to Cuban, American, and regional economic development interests. This is especially relevant in light of growing concerns about the region's energy infrastructure—in particular, the oil and gas industries of Mexico and Venezuela, where there is growing evidence that policy priorities in both countries might be hindering their capacity to deliver on their contractual obligations to export oil to the U.S. market.

Drawing Conclusions

The final component of the analysis will draw conclusions from the previous sections, in addition to the contributions of the research team, and will provide concrete policy recommendations on the challenges and opportunities facing American policymakers hoping to engage Cuba in terms of energy and infrastructure investment, development, and revitalization. This analysis will allow us to discuss the following questions germane to the discussions of Cuba's energy future and the related U.S. need to articulate a new vision of how to manage international energy interdependence most effectively. Cuba's centrality to that discussion led us to ask the following questions:

—In what way will the ongoing development and evolution of Cupet, Cuba's state oil company, limit or obstruct U.S. efforts to meet its strategic objectives? This includes the relationship between Cupet and Petróleos de Venezuela, as well as the Venezuelan state.

—What role can international oil companies play in the development of energy resources and infrastructure in Cuba in both the near and long term? Cuba is seeking to develop the production capability of its North Coast Reserves, and national oil companies (NOCs) from nine different countries have signed lease agreements with the Cuban regime for offshore tracts.

—What impact will competition from Brazil, Russia, India, and China over scarce petroleum resources have on U.S. energy security? Recently Russia and Cuba have signed an energy-development agreement, and the Chinese have made incursions into Latin American energy markets.

Answering these questions will allow us to develop a refined set of policy recommendations that enhance the prospects for cooperation and perhaps for the amelioration of relations between these neighbors.

Notes

1. Russell Blinch, "Cuba's Offshore Oil Hopes Rise, U.S. Role Uncertain," Reuters News Service, June 15, 2009 (www.reuters.com/article/globalNews/idUSTRE55E67520090615).

2. Carlos Pascual, "The Geopolitics of Energy: From Security to Survival," (Brookings Institution, January 2008) (www.brookings.edu/papers/2008/01_energy_pascual.aspx).

3. Amy Myers Jaffe, "The Changing Role of National Oil Companies in International Energy Markets—Introduction and Summary Conclusions," PowerPoint presentation, March 1, 2007, p. 6 (www.rice.edu/energy/publications/docs/NOCs/Presentations/Hou-Jaffe-KeyFindings.pdf).

4. U.S. Geological Survey, "Assessment of Undiscovered Oil and Gas Reserves of the North Cuba Basin, Cuba, 2004," Fact Sheet 2005-3009 (Washington: USGS, February 2005), p. 2.

5. Jeremy Rifkin, *The Empathic Civilization: The Race to Global Consciousness in a World in Crisis* (New York: Penguin Books, 2009), pp. 518–19, 553.

6. Partnership for the Americas Commission, *Rethinking U.S.-Latin American Relations,* final report (Washington, D.C.: Brookings Institution, November 2008), p. 28.

7. Brookings Project on U.S. Policy toward a Cuba in Transition, *Cuba: A New Policy of Critical and Constructive Engagement* (Washington, D.C.: Brookings Institution, April 2009), pp. 3–4, 15.

8. Ibid.

9. Jeff Franks, "Cuba Oil Claims Raise Eyebrows in Energy World," Reuters, October 24, 2008 (www.reuters.com/article/internalReutersGenNews/idUSN 2347317820081024). See also CubaNews, "Petrobras Inks E&P Deal with Cuba," October 31, 2008 (www.rigzone/news/article.asp?a_id=68675).

10. Partnership for the Americas Commission, *Rethinking U.S.-Latin American Relations,* p. 29.

11. Ibid., p. 29.

12. See Juan A. B. Belt, "Power Sector Reforms in Market and Transition Economies: Lessons for Cuba," paper presented at the sixteenth annual meeting of the Association for the Study of the Cuban Economy, Miami (August 3, 2006), *Cuba in Transition* 17 (2008): 75–89 (http://pdf.usaid.gov/pdf_docs/PNADG827.pdf); Jonathan Benjamin-Alvarado, "Commentary on 'Cuba's Energy Challenge: A Second Look,' by Piñón-Cervera," *Cuba in Transition* 15 (2006): 124–26; Jorge Hernandez Fonseca, "El Programa Brasileña de Etanol: Lecciones para Cuba" [The Brazilian ethanol program: Lessons for Cuba], *Cuba in Transition* 17 (2008): 206–11; Jorge R. Piñón-Cervera, "Cuba's Energy Challenge: A Second Look," *Cuba in Transition* 15 (2006): 110–23.

13. Fidel Castro Diaz-Balart, *Energia Nuclear: ¿Peligro Ambiental o Solucion para el Siglo XXI?* [Nuclear energy: Environmental danger or solution for the 21st century?] (Turin, Italy: Ediciones Mec Grafic, 1997); Juan Jardon, *Energia y Medio Ambiente: Una Perspectiva Económico-Social* [Energy and the environment: A socioeconomic perspective] (Mexico City: Plaza y Valdes Editores, 1995); Tirso Saenz and Emilio G. Capote, *Ciencia y Tecnologia en Cuba* [Science and technology in Cuba] (Havana: Editorial de Ciencias Sociales, 1989).

14. Alternativa Bolivariana para las Americas is a Latin American trade association designed as an alternative to the Free Trade of the Americas Agreement. Petrocaribe S.A. is a Caribbean oil alliance of Cuba and fourteen other Caribbean states with Venezuela, launched in June 2005 to purchase oil on conditions of preferential payment. The payment system allows for a few nations to buy oil on favorable terms: they pay market value but only a certain amount of money is needed upfront; they can pay the remainder through a twenty-five-year financing agreement at 1 percent interest.

15. Drexel Kleber, "The U.S. Department of Defense: Valuing Energy Security," *Journal of Energy Security* (June 2009): 3 (www.ensec.org).

16. Ibid., p. 4.

17. Mahmoud Amin El-Gamal and Amy Myers Jaffe, "Energy, Financial Contagion, and the Dollar," Working Paper Series, The Global Energy Market: Comprehensive Strategies to Meet Geopolitical and Financial Risks (Houston: Rice University, James A. Baker III Institute for Public Policy, 2008), p. 26.

18. Ibid., pp. 28–29.

19. Mexico's Pemex and Venezuela's PDVSA are two of the largest suppliers of oil to the United States, after Canada and Saudi Arabia. Mexico is third, with 1,433,000 barrels per day, and Venezuela is fourth, with 1,162,000 barrels per day.

20. See Joe Barnes and Matthew E. Chen, "NOCs and U.S. Foreign Policy," paper (Houston: James A. Baker III Institute for Public Policy and the Japan Petroleum Center, Rice University, 2007), p. 4 (www.rice.edu/energy/publications/docs/NOCs/Papers/NOC_US-ForeignPolicy_Barnes-Chen-revised.pdf).

21. Simon Romero, "Venezuela Suspends Heating Aid to the U.S.," *New York Times,* January 6, 2009.

22. U.S. Energy Information Administration, "Crude Oil and Total Petroleum Imports, Top 15 Countries," April 29, 2010 (www.eia.doe.gov/pub/oil_gas/petroleum/data_publications/company_level_imports/current/import.html).

23. Jaffe, "Changing Role of National Oil Companies," p. 6. See also Amy Myers Jaffe and Ronald Soligo, "The International Oil Companies," paper (Houston: Rice University, James A. Baker III Institute for Public Policy, 2007), p. 6.

24. U.S. Energy Information Administration, "Crude Oil and Total Petroleum Imports, Top 15 Countries."

25. Ibid., p. 10.

26. Flynt Leverett and Jeffrey Bader, "Managing China-U.S. Energy Competition in the Middle East," *Washington Quarterly* 29 (Winter 2005–06): 197.

27. U.S. Geological Survey, "Assessment of Undiscovered Oil and Gas Resources of the North Cuba Basin, Cuba, 2004," World Assessment of Oil and Gas Fact Sheet (http://pubs.usgs.gov/fs/2005/3009/pdf/fs2005_3009.pdf).

28. See Jonathan Benjamin-Alvarado, *Power to the People: Energy and the Cuban Nuclear Program* (New York: Routledge, 2000).

two

Extracting Cuba's Oil and Gas: Challenges and Opportunities

JORGE R. PIÑÓN AND JONATHAN BENJAMIN-ALVARADO

Now . . . the possibility of Cuba's harboring a great oil reservoir is again under investigation.

"Cuban Dream," *Time* magazine, August 15, 1938

"Oil Exploration Pushed in Cuba as Major Companies Get Rights"

New York Times, January 8, 1958

"Why Cuban Dreams of Major Oil Discoveries Might Come True"

U.S. News & World Report, March 4, 2009

Potential offshore oil deposits were identified in Cuban waters in 2004, yet amid widespread speculation as to their magnitude and potential, there has been little exploratory work undertaken to evaluate the actual amount of oil and gas reserves that are present, the feasibility of extraction, and the deposits' eventual productive capacity. In the recent as in the more distant past, there have been hints that there may be an oil giant in the Florida Strait ripe for investment from its neighbors to the north. Insofar as Cuba remains off-limits to American firms, the notion becomes all the more alluring and heightens the sense of possibility: What might a United States–Cuba oil partnership lead to?

The primary objective of this chapter is to provide an analysis of the upstream oil and natural gas potential in Cuba, in the following areas:

—Actual potential hydrocarbon (crude oil and natural gas) production figures

—Potential realized oil prices from joint-venture projects

—The possible role of Cuba's offshore Exclusive Economic Zone (EEZ)

—Types of concessions that could be granted in the Cuban EEZ

—An assessment of the future upstream oil and natural gas development challenges

These analyses should provide ample basis for the broader discussion of the challenges in the electrical power sector in Cuba, energy balances, and the potential for biofuels discussed in chapters 3 and 4 of this volume. In the appendix to this chapter are tables of data relevant to Cuban energy development (the Cuban supply/demand balance; Cuban gasoline prices; Cuban lubricants–motor oil production figures; Cuban refinery production figures; Cuban estimated annual petroleum import values; Cuban-Venezuelan estimated petroleum debt).

All told, the data and analysis presented in this chapter are designed to serve as the basis of discussions of the broader implications of Cuban energy development: both the direct challenges implied for Cuban policymakers and the possible opportunities that these challenges present to potential American partners and policymakers as they relate to energy cooperation and energy security concerns.

Onshore and Coastal Oil Resources

Since early colonial days we find reports of Spanish explorers and *conquistadores* waterproofing the hulls of their galleons with bitumen-tar-like material found in the vicinity of Havana and Cardenas harbors in eastern Cuba. The first major Cuban field to be discovered, in 1881, was the shallow Motembo oil field in central Las Villas province, consisting of light condensates.[1]

Even though expectations ran high during the first half of the twentieth century for the discovery of more oil fields in Cuba, exploration results were disappointing, and only several small oil fields were discovered. The situation changed during the 1970s, when, with Soviet assistance, the Varadero oil field, on the coast east of Havana, was discovered in 1971.[2]

Foreign Participation in Cuban Oil

In 1993, after the fall of the Soviet Union in 1991, Cuba opened its oil and gas exploration and production sector to foreign oil companies. Thirty-three onshore and coastal blocks were offered during its first international round of bidding (see figure 2-1).[3] ("Onshore" means on dry land; "coastal" refers to coastal waters—from the water's edge to the coastal shelf; "offshore" refers primarily to deepwater resources—beyond the coastal shelf.) In order to attract foreign oil companies to explore and produce Cuba's potential

Figure 2-1. Cuba's Land and Marine Oil Blocks, 2009

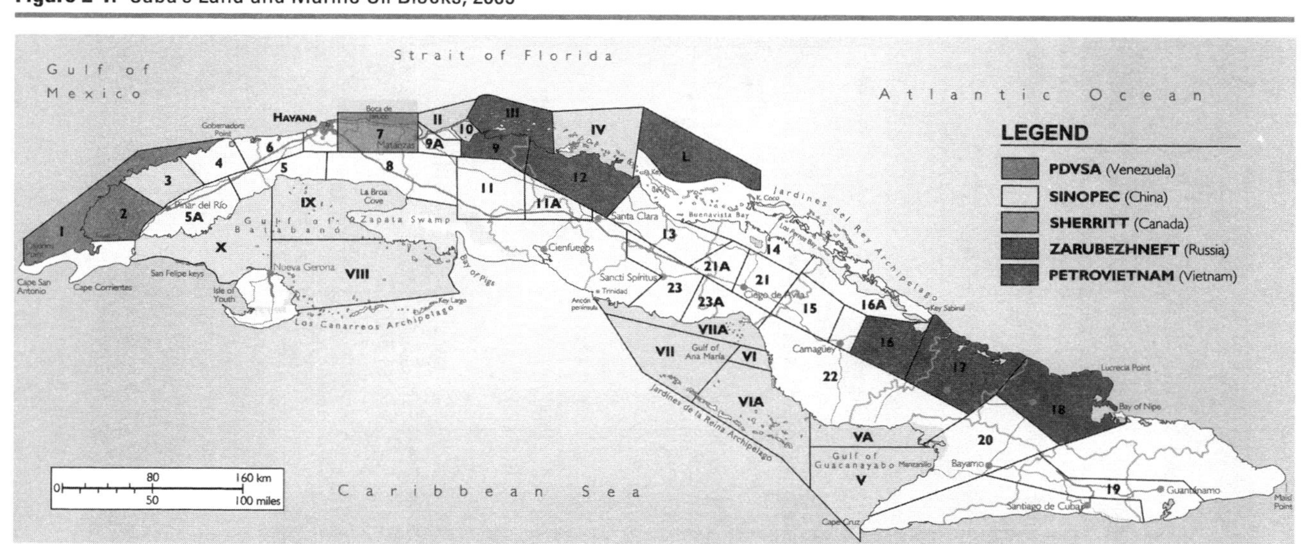

Source: Jorge R. Piñón.

hydrocarbon resources, the Cuban government, through Unión Cubapetróleo S.A. (Cupet), the state oil company managed by the Ministry of Basic Industry (Minbas), adopted a contractual format known as a production-sharing agreement (PSA), an arrangement used by many countries and generally accepted by major international oil companies.[4]

Legal Status of Oil Reserves

All subsoil hydrocarbon reserves are the property of the state and are under the ultimate control of the government of Cuba. Under its PSA, Cupet awards the rights to a third-party contractor to explore and produce hydrocarbons within a specific geographical area at its own risk. The third-party contractor is the joint-venture enterprise established between the Cuban government entity and the foreign investor. The contractor is responsible for supplying all capital, equipment, installations, technology, and personnel needed to carry out the operations as outlined in the contract.

PSAs are generally divided into exploration and production phases, each phase having its own set of performance requirements. If no exploratory work is conducted within the specified exploration period, typically three to seven years, or insufficient quantities of crude oil or natural gas are found and the reservoir is declared uncommercial, the concession can be withdrawn and the contractor will be unable to recover any of its capital investments. If the reservoir is considered commercially viable, then the first oil extracted from the concession is allocated to the contractor to recover its capital investment and other exploration costs—called "cost oil"—with a limit on what percentage of production can be allocated as cost oil.

Once costs have been recovered, the remaining oil—called "profit oil"—is divided between Cupet and the contractor in agreed proportions as outlined in the PSA. Exploration and production terms typically have a duration of twenty-five to thirty years. The contractor generally pays taxes of 25 percent on personnel salaries and 30 percent on net profits. The contractor is also allowed to dispose of its share of production by exporting it in kind or selling it to Cupet according to an agreed-upon price formula.

Under the current PSA terms and conditions, the foreign partner does not receive any allocation of associated natural gas production; however, all costs related to such production are recoverable.

In most extant PSAs throughout the world today, the national oil company also has the option to participate as a joint-venture partner in the project, by providing its percentage share of capital investment and receiving its propor-

Table 2-1. Cuba's Hydrocarbon Production, 2003 to 2008

Year	Crude oil and natural gas liquids (barrels per day)	Natural gas (barrels of oil per day equivalent)	Year	Crude oil and natural gas liquids (barrels per day)	Natural gas (barrels of oil per day equivalent)
2003	65,531	11,339	2006	51,644	18,794
2004	57,930	12,135	2007	51,733	20,988
2005	52,269	12,808	2008	51,834	20,142

Source: Author compilation, based on data published by the Oficina Nacional de Estadísticas de Cuba, 2009.

tional share of both cost oil and profit oil. In most cases the most experienced partner remains as the operator of the project.

Since 1991 Cuba has seen close to $2 billion of direct foreign investment spent in its upstream oil and natural gas sector, with very good results. Crude oil liquids production reached a peak level of 65,531 b/d in 2003, up from 9,090 b/d in 1991. Between 2003 and 2010 Cuba saw its crude oil production level out at around 52,000 b/d.[5]

Characteristics of Cuba's Oil Deposits

Most hydrocarbon discoveries in Cuba have been heavy, sour-quality crude oil, along with associated natural gas found in fractured carbonate reservoirs of Jurassic and pre–Upper Cretaceous Campanian plays in the North Cuba Fold and Thrust Belt, along a 200-by-15-kilometer stretch of the northern coastal and onshore region between Havana and Corralillo.[6] Cuba's heavy oil recovery rates are around 7 percent of proven reserves,[7] estimated by the U.S. Energy Information Administration to be at 750 million barrels. These low recovery factors are due to the viscous quality of the crude oil and the nature of the geology.

The majority of the production from the Varadero, Puerto Escondido, and Boca de Jaruco fields is 9- to 12-degree API gravity heavy crude oil with a high content of sulfur and heavy metals.[8] Due to the low quality of crude oil currently being produced in Cuba and the final end-use of the same as fuel for Cuban electric power plants, the price basis for the island's production is a discounted price off U.S. Gulf Coast No. 6 industrial fuel oil. Cuba's realized crude oil value could substantially improve once it is able either to monetize its heavy oil production in its own future heavy oil conversion refinery processing capacity, or to market its crude oil to U.S. Gulf Coast refining companies.

Table 2-2. Per-Barrel Prices of Three Grades of Cuban Oil Sold by Sherritt Inc., and Realized Price, 2003 to 2008
U.S.$

Year	West Texas Intermediate	Gulf Coast #6 fuel oil	Realized	Less than 20-degree crude
2003	31.08	23.79	27.36	21.88
2004	41.51	24.47	25.98	28.95
2005	56.64	35.9	35.56	40.54
2006	66.05	45.33	41.51	50.19
2007	72.34	52.85	42.53	56.69
2008	99.67	72.63	55.88	81.25

Source: Author compilation, based on Sherritt Inc.'s financial reports, 2003–08, and U.S. Energy Information Administration independent statistics and analysis (www.eia.doe.gov).

Most of Cuba's heavy oil production today is the result of the PSA between Cupet and Canada's Sherritt International, using directional drilling technology from onshore locations targeting coastal offshore reservoirs one to five kilometers from shore. Sherritt reported in its 2008 financial statements that gross working interest production amounted to 31,233 b/d, which represented 60 percent of the total Cuban production. Production costs of $7.80 per barrel from a realized price of $55.88 per barrel represented an annual gross profit earnings before interest, taxes, depreciation, and amortization (EBITDA) of $206 million.[9]

Another Canadian oil company active in Cuba for many years was Montreal-based Pebercan, which in 2009, after a payment dispute with Cupet, agreed to terminate all of its contractual obligations on the island.[10] The decision will financially affect Sherritt in future years, as Sherritt had a number of producing properties in association with Pebercan. Going back to 2007, Sherritt became the de facto operator of Pebercan's contractual obligations in Cuba, essentially relegating Pebercan to the role of silent financial partner in terms of oil exploration and production on the island. In 2008, Cuba's own financial crisis was brought about by the precipitous drop in price of Cuban nickel exports, and hurricane-related problems in the domestic agricultural market increased the need to boost food imports, forcing Cupet to delay its payments. By the beginning of 2009, there was no realistic reason for Pebercan to renegotiate its contracts for exploration, largely because it no longer held the technical role in the exploration for oil, nor did it have a stake in any other oil industry venue on the island. Cuba's cash crisis had subsided sufficiently so that Cuba was able to come to an agreement on the repayment of the existing obligations, effectively ending Pebercan's presence in Cuba.

Sherritt and Cupet are currently negotiating contractual terms and conditions in order to implement secondary enhanced recovery and production projects in the Varadero oil field, which could substantially increase current production levels.[11] Heavy oil reservoirs generally have low primary recovery rates of less than 10 percent. Long field life and stable and predictable production, along with new enhanced recovery technologies, make heavy oil development a more attractive investment today than in previous decades, at current world oil prices of over $55 per barrel.

Secondary and tertiary enhanced oil recovery (EOR) technologies such as water or steam injection, natural gas reinjection, and carbon dioxide injection can increase recovery rates by 15 to 20 percent, depending on the permeability of the rocks and the viscosity of the oil.[12]

As of 2010, Sherritt has concessions to block 7, and four national (state-owned) oil companies also hold onshore concessions. Venezuela's PDVSA (Petróleos de Venezuela S.A.) has the concessionary rights to the westernmost blocks, 1 and 2, in Pinar del Rio province, which it acquired in January of 2007. Also in western Pinar del Rio province, China's Sinopec (China Petroleum and Chemical Corporation) acquired the exploratory rights to onshore block 3 in February 2005. In the central part of the island, Russia's Zarubezhneft was awarded in November 2009 exploratory concessions to blocks L, III, 9 and 12. In eastern Cuba, PetroVietnam holds concessions in blocks 16, 17, and 18 (see figure 2-1).[13]

As of spring 2010, the state-owned firms have conducted only seismic work in their blocks. In 2005, a Sinopec subsidiary, Great Wall Drilling Co., signed a service contract with Cupet to provide a number of bidirectional drilling rigs and other service equipment. A bidirectional rig allows the oil company to drill down vertically and then horizontally or at an angle to access oil reserves. This makes it possible for oil rigs to be placed onshore to access reserves relatively near the shore in coastal reserves. They are currently leased by Sherritt and Cupet and are operating in block 7.[14] This activity has been the source of recent rumors and speculation in the United States that China was actively drilling in Gulf of Mexico waters off the Florida coast.[15] Cupet is currently conducting exploratory work without foreign assistance east of Havana in prospect areas such as Guanabo, Tarara, Santa Maria, and others.[16] Cuba's onshore and coastal (in coastal waters) heavy oil production seems to have reached a plateau at around 52,000 b/d; but once Cupet can gain access to the services, technology, equipment, and capital available through independent U.S. oil and oil services and equipment companies, Cuba's onshore and coastal heavy oil production potential could grow to an amount in excess of 75,000 b/d. Cuba will be able to realize these gains by implementing enhanced secondary recovery methods. If Cuba can access advanced technology (primarily U.S.-based technology) and more investment capital, there is a reasonable expectation that Cuba could increase its recovery rate to somewhere between 17 and 20 percent. This is the recovery factor for similar heavy crude fields in the United States and Canada.

Natural Gas Resources

All of Cuba's natural gas production is associated natural gas—found within—the crude oil reservoirs. To date the island's geology has not proved to be a major source of dry, non-associated natural gas reservoirs. Annual production is averaging 43 billon cubic feet (bcf) or 21,000 barrels of oil per day equivalent (boe).[17] According to the U.S. Energy Information Administration, there are estimated reserves of 2,500 trillion cubic feet (tcf); according to Cupet officials current natural gas recovery rates are approximately 94 percent.

Cuba's associated-gas production is a true success story. For many years the gas was "flared," creating considerable air and visual pollution in the tourist-sensitive area along the Via Blanca highway as it approaches the beach resort area of Varadero. Economic incentives allowed Cupet to develop a business solution to the "rotten egg" smell with Sherritt. Locally produced associated natural gas from the Varadero, Boca de Jaruco, and Puerto Escondido fields is now being used as fuel for on-site power generating plants of 400 MW total capacity.[18]

The power plants and related processing units for "sour" gas are owned and operated by Energas, a joint venture in which Sherritt has a one-third indirect interest; Cupet, which supplies gas at no cost to the joint venture, has a one-third interest; and Unión Eléctrica, which buys all the power from the plants, also has a one-third interest.

Cupet also built a system of pipelines from the Puerto Escondido and Boca de Jaruco fields, which transports natural gas to the Santa Cruz del Norte thermoelectric power plant and to the city of Havana, and crude oil to the oil superport in Matanzas and the Ñico Lopez refinery in Havana.[19]

Cuba will probably have little choice but to develop an energy policy that relies heavily on clean-burning natural gas as its fuel of choice for electrical

Table 2-3. Cuba's Crude Oil Production, 2003 to 2008
Metric tons

Production unit	2003	2004	2005	2006	2007	2008
EPEP—Occidente[a]	1,902.2	1,514.4	1,545.8	1,737.7	1,865.0	1,813.2
EPEP—Centro[b]	1,735.1	1,701.9	1,354.0	1,129.2	1,009.6	1,161.6
Diluents[c]	42.5	36.7	35.3	33.1	30.4	28.3
Total production (metric tons)	3,679.8	3,253.0	2,935.1	2,900.0	2,905.0	3,003.1
Total production (barrels per day)[d]	64,018	56,593	51,063	50,452	50,539	52,246

Source: Author compilation, based on statistics compiled by Oficina Nacional de Estadísticas de Cuba, 2008.

a. Empresa de Perforación y Extracción de Petróleo–Occidente, Cupet's business unit responsible for the exploration and production activities in La Habana and Pinar del Rio provinces, including the production areas of Santa Cruz del Norte, Boca de Jaruco, and Puerto Escondido.

b. Empresa de Perforación y Extracción de Petróleo–Centro, Cupet's business unit responsible for the exploration and production activities in Matanzas and other central provinces; core activities are located in the Varadero-Cardenas area.

c. Cuban crude oil is too viscous to be easily pumped or transported by pipeline and/or marine vessel. Typically naphtha, condensate, or other diluents are added to improve the viscosity of the crude oil.

d. Metric tons converted to barrels per day. The correct factor to use for converting metric tons of oil to barrels per day (how many barrels per ton) is determined by the API gravity of the crude oil stream. Cuban crude oil API gravity varies from a low of 8 to 10 degrees (Varadero) to a high of 12 to 15 (Seboruco, Canasi, Puerto Escondido). Owing to production decline of the Varadero field we have selected a weighted average conversion factor of 6.35 barrels per ton.

power generation. Drivers of this necessity are the inevitable rationalization of the oil-refining industry in Cuba (because of its outdated technology, which is unable to process heavy crude oil), and the country's environmentally sensitive tourist industry. Cuba's future natural gas needs could be filled by importing liquefied natural gas (LNG) from Trinidad and Tobago, which Puerto Rico and the Dominican Republic are currently doing, or by future Venezuelan production. A regasification facility to receive Venezuelan sourced liquid natural gas is currently being planned for the southern-coast port city of Cienfuegos by Venezuela's PDVSA and Cupet. Two one-million-ton regasification trains are planned for 2012, at a cost of over $400 million. The natural gas is destined as fuel for that city's thermoelectric power plant, local industry, and future petrochemical plants.[20]

Cuba's Deep Water: The Exclusive Economic Zone

The future of Cuba's oil and gas exploration and production sector could very well be in the deep offshore Gulf of Mexico waters along the western approaches to the Florida Strait and the eastern extension of Mexico's

Figure 2-2. Cuba's Exclusive Economic Zone (EEZ)

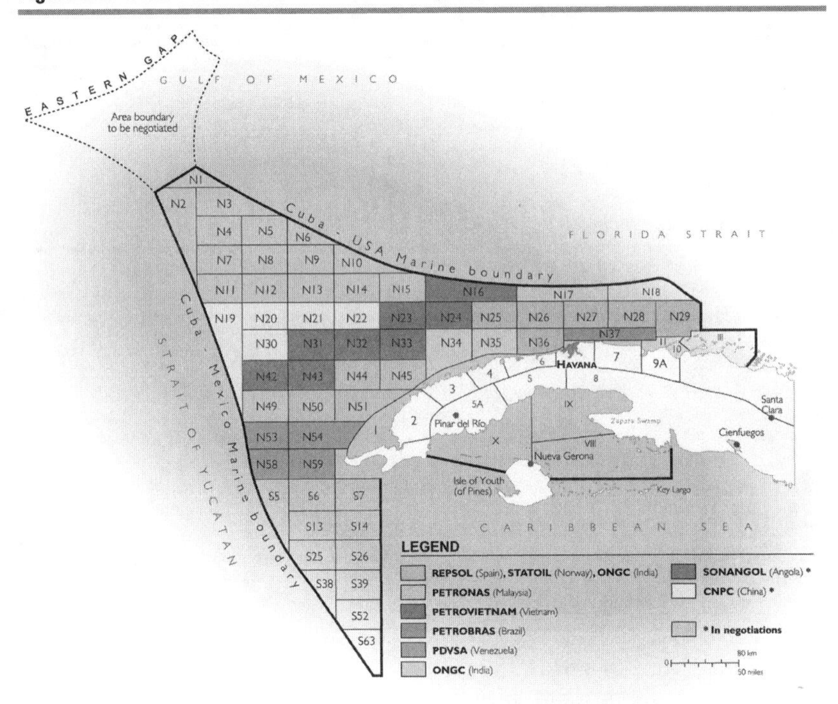

Source: Jorge R. Piñón.

Yucatán Peninsula. Cuba's Exclusive Economic Zone (EEZ) in the Gulf of Mexico is a 46,000-square-mile area that Cupet has divided into fifty-nine exploration blocks of approximately 772 square miles each. The average ocean depth is 6,500 feet, but some blocks are as deep as 13,000 feet.[21]

Geography of Oil in the Gulf of Mexico

The EEZ lies between Mexico, Cuba, and the United States, within demarcation boundaries agreed to in 1977. The northernmost of the blocks lies south of the Dry Tortugas, off Florida's southwest coast. The northwesternmost blocks are situated next to the Gulf of Mexico's eastern gap, a sizable portion of the eastern Gulf, west of the Florida EEZ and north of the Cuban EEZ, for which economic exclusivity rights have not been negotiated, and 100 kilometers from the southernmost limit of acreage, offered as lease 181 by the U.S. Mineral Management Services, on the outer continental shelf off Florida's west coast.[22]

Although the maritime boundary agreement between Cuba and the United States has been submitted to the U.S. Senate, for political reasons—not because of any objection in the boundary itself—it has not been ratified by that body. Cuba and the United States have since agreed to provisional application of the agreement, pending ratification, by exchanging agreement notes every two years that extend the provisional application of the agreement. The demarcation of the Gulf of Mexico's eastern gap itself, which will include Cuba, Mexico, and the United States, is still open for negotiation, and awaits improvements in the diplomatic relations between Washington and Havana.

A February 2005 U.S. Geological Survey report, "Assessment of Undiscovered Oil and Gas Resources of the North Cuba Basin 2004," estimates a mean of 4.6 billion barrels of undiscovered oil and a mean of 9.8 trillion cubic feet of undiscovered natural gas along Cuba's North Belt Thrust. The high-end potential of the North Cuba Basin could be 9.3 billion barrels of undiscovered oil and of 21.8 trillion cubic feet of undiscovered natural gas, according to the report.[23] If these undiscovered reserves are certified as recoverable, they will rank Cuba among major Latin American oil producers and exporters such as Colombia and Ecuador.

Industry experts have categorized Cuba's EEZ as high risk from the technical geosciences standpoint—there might not be any oil or gas there—but some reports indicate that some hydrocarbon potential might exist. Meanwhile, Cuban government sources estimate the potential of the whole EEZ at an optimistic 20 billion barrels of undiscovered reserves.[24] This figure includes the 5 billion barrels that the U.S. Geological Survey estimates in the Cuba North Belt Thrust, and an additional 15 billion barrels of undiscovered reserves in the North Cuba Foreland Basin, the Florida and Campeche escarpments, on the shelf margin of the Florida Platform, and in the Gulf of Mexico Sigsbee Basin. Very little seismic work and exploratory drilling have been done outside of North Cuba's Fold and Thrust Belt, the North Cuba Foreland Basin, and the U.S. Geological Survey's Florida Platform Margin Carbonate assessments units (AUs).[25] This can be interpreted as meaning that there is a high likelihood of oil and gas in Cuba's offshore reserves. Moreover, a basic analysis of the geological formations by Cuban analysts suggest that the potential for additional reserves is likely.

In most experts' opinion, a lot of exploratory work has yet to be done to substantiate the high-end estimates put forth by Cuban geologists, regardless of the technical soundness of the data presented in support of their estimate.

Table 2-4. Cuba's Concessions in Its Exclusive Economic Zone, 2009

Block numbers	Year awarded	Company (country)	Consignee equity (percent)
N25–29, N36	2002	Repsol (Spain)	40
	2005	Statoil–Norsk Hydro (Norway)	30
	2005	ONGC (India)	30
N34, N35	2006	ONGC (India)	100
N44, N45	2006	Petronas (Malaysia)	100
N53, N54, NN58, NN59	2007	PDVSA (Venezuela)	100
N31, N32, N42, N43	2007	PetroVietnam	100
N37	2008	Petrobras (Brazil)	100

Source: Author compilation, based on data published by Oficina Nacional de Estadísticas de Cuba, 2009.

As of late 2009, Cupet had consigned twenty-one of the fifty-nine deepwater blocks in Cuba's EEZ to seven international oil companies. Sherritt lost its offshore concession for blocks N16, 23–24, and 33 in 2007, after failing to meet the contractual requirements of the seven-year exploratory term of the company's PSA with Cupet. This event did not affect its onshore and coastal concessions for block 7, however.[26] As of early 2010 Cupet was in negotiations with China's CNOOC (Chinese National Offshore Oil Corporation) and Angola's Sonangol, both national oil companies, for exploratory concessions for a series of nine deepwater Gulf of Mexico blocks.

Exploratory Drilling So Far

In June 2004 Spain's Repsol-YPF drilled the first exploratory well within Cuba's EEZ, Yamagua #1, located in block 27, about twenty miles northeast of Havana and about ninety-five miles southwest of Key West. The well reached a depth of 10,819 feet in water depths of 1,500 feet and is estimated to have cost over $40 million. According to press reports of July 29, 2004, Repsol's chief operating officer, Ramon Blanco, said that the drilling results were promising: "The existence of a petroleum system has been confirmed. Also we have been able to prove the presence of high-quality reserves. Nevertheless, the well has been considered noncommercial and at this stage the group is defining future exploration activities in the area."[27] Repsol's CEO, Antonio Brufau, announced in Madrid on May 31, 2005, the company's commitment to additional exploration efforts in Cuba. He also announced that the Norwegian oil giant Norsk Hydro and India's ONGC (Oil and Natural Gas Corporation), India's biggest energy exploration company, would be project partners.[28] The participation of Norsk Hydro was an indication of the importance and potential of the project, because Norsk Hydro is recognized in industry circles for its deepwater exploration and operational expertise.

Repsol is expected to drill a second exploratory well in 2010 or by early 2011, as it was able to secure a deepwater, semi-submersible rig that will not trigger U.S. sanctions, as set forth in the Helms-Burton Act. According to the act, any U.S. patented machinery or technology that contains more than 10 percent of U.S. components is considered to be in violation of U.S. trade sanctions against Cuba.

According to press reports, Repsol has leased the semi-submersible Scarabeo 9 from Italy's Eni S.p.A. This state-of-the-art semi-submersible, built without any U.S. components in China, could be in Cuban waters by the end of the year. In the meantime, Repsol was able to negotiate an extension on the seven-year exploratory terms of its 2002 PSA.[29]

As discussed earlier, this type of deepwater exploration is expensive and carries high geological and technical risk, risks that companies such as Repsol-YPF, Statoil–Norsk Hydro (Statoil and the oil and gas business of Norsk Hydro merged in 2007), and Brazil's NOC, Petrobras, certainly have the necessary deepwater expertise to handle. Repsol recently announced a new deepwater oil discovery at the Buckskin Prospect, located at approximately 6,920 feet of water in the U.S. Gulf of Mexico. A well was drilled to an unprecedented depth of 29,404 feet, demonstrating the Spanish oil company's technical and operational competencies.[30]

A further consideration for most international oil companies is that an anticipated crude oil price of more than $65 per barrel would need to be in place before any deepwater projects would be financially worthwhile. Today, such prices seem to be the case. If successful, the Cuba North Coast deepwater project would take from two to three years to bring into full development, at an estimated total cost of $1 billion to $3 billion.

Environmental Issues

Some critics have raised the specter of a looming environmental disaster should Cuba proceed with plans to begin extensive drilling operations in its EEZ.[31] They fear that Cuba's efforts to extract oil from its deepwater reserves will result in drilling rig accidents and shipping disasters that recall the environmental catastrophes of Mexico's 1979 Pemex Ixtoc I well blowout, or the *Exxon Valdez* oil spill in Prince William Sound of Alaska in 1989, not to mention the *Deepwater Horizon* disaster of 2010, ongoing at this writing. But oil exploration is an inherently risky enterprise. There is always a balance between the positive and negative trade-offs of energy security and environmental integrity. All of the relevant actors on the Cuban scene—Repsol, Statoil–Norsk Hydro, and Petrobras—are already working in the Gulf of Mexico operating platforms under contract to international oil companies based in the United States.

What international oil companies operating in Cuba lack is immediate access to U.S. resources and technology needed to prevent or mitigate a catastrophic incident in Cuban waters similar to that of the *Deepwater Horizon.* The United States administration should review the regulations that prohibit the transfer of technology, equipment, or personnel to international oil companies operating in Cuba in the case of an oil-related emergency. Executive orders should be in place that would allow this transfer along with protocols and emergency plans in cooperation with Cuban authorities.

Future Upstream Oil and Natural Gas Development Challenges

Today, a number of key preconditions must be positive in order for a major oil company to engage anywhere in the world in the exploration and production of a new, unexplored hydrocarbon frontier: geology, capital, technology, know-how, and the ability to monetize the discovered resources, if any. Anyone who evaluates current and reported future international oil companies that are involved in Cuba's deepwater search for oil and gas must understand their competency, strategic objectives, and their possible long-term contribution to the island's efforts to become energy independent.[32]

As long as the economic and trade restrictions imposed on the government of Cuba by the U.S. government continue, all companies, regardless of their technical competence, will have a very difficult time in monetizing any newly discovered hydrocarbon resources because they need access to the U.S. oil services and equipment market. Also Cuba urgently needs, but does not have, a complex oil-refining system able to process the probable large quantities of heavy crude oil found in Cuba's offshore waters. Until Cuba develops its own heavy-oil-refining infrastructure, any newly found oil most likely will have to be exported. Its natural market is the United States, the largest importer of oil in the world—yet that market is closed to Cuba by the trade and commercial restrictions currently in place.

"The good news is we found oil; the bad news is we found oil" will be the likely announcement of any new oil discovery in the corporate headquarters of the oil companies doing business in Cuba. Repsol and Statoil–Norsk Hydro certainly have the necessary capacities and competencies described earlier to develop and produce any oil they find. Their challenge is how and where to commercialize the "black gold"—refine it and bring it to market. Some international oil companies are in Cuba for strictly economic and business reasons. Others acquire concessions in the expectations that U.S. policy will change before the end of their seven-year exploratory term, at which time they will be able to bring in a majority U.S. oil company as a partner. Others could be grandstanding on behalf of the Cuban government: putting a spotlight on Cuba's oil and natural gas potential in order to influence United States special interest groups to lobby for lifting the economic and trade restrictions.

Spain's Repsol is the only privately held oil company active in Cuba's EEZ today. All others are NOCs, oil companies whose controlling shares are owned by their respective national governments. That said, Norway's Statoil–Norsk Hydro and Brazil's Petrobras, even though their respective governments have large equity holdings, behave and act as true privately held international oil companies with no national political agenda. These two companies also have the capital resources necessary to undertake large and expensive projects on their own.

Norsk Hydro was the world's fourth-largest integrated aluminum company in the world, and was investing heavily in oil, gas, and renewable energy. In 2007 it merged its energy operations with StatoilHydro of Norway, its chief competitor in the North Sea. Petrobras is the largest oil company in Latin America by market capitalization and revenue, the largest company headquartered in the Southern Hemisphere, and the eighth-largest in the world in terms of revenue. It is a significant oil producer, with output of more than 2 million barrels of oil equivalent per day, as well as a major distributor of oil products. The company also owns oil refineries and oil tankers. Petrobras is a world leader in development of advanced technology for deepwater and ultra-deepwater oil production. Because Statoil–Norsk Hydro and Petrobras no longer are state-owned monopolies, they are not dependent on their respective governments' public treasuries for capital, alleviating some pressure on the countries' national budgets. Instead, their sources of capital come directly from private and public investors and partners.

India's ONGC and Malaysia's Petronas also have the technical expertise for offshore deepwater exploration and production, but their capacity to raise capital for high-risk, high-reward projects in areas outside their nations' geographical influence is limited. PetroVietnam is probably the weakest of all national oil companies involved in Cuba today. Although it possesses the requisite deepwater operational know-how, it cannot raise sufficient investment capital for high-risk projects outside Vietnam within its national budget limitations. Because it is a national oil company, most if not all of its investment capital is drawn from the government, and the Vietnamese government currently does not possess the cash reserves to bankroll the highly capital-intensive costs of oil exploration in Cuba.

Venezuela's PDVSA has no deepwater expertise but does have the capital to fund its Cuba ventures if it chooses. In addition, the role PDVSA plays in Cuba is highly political, so one can safely assume that a number of PDVSA's joint-venture projects with Cupet will most likely be funded, even if they lack strategic purpose or an acceptable economic rate of return for PDVSA.[33] Of the estimated 2003–09 value of $14 billion for Venezuela for oil imports, over $8 billion is being offset by the thousands of Cuban doctors and other professionals being deployed across Venezuela, while the balance of nearly $6 billion has been converted into long-term (twenty-five-year) debt.[34] This represents a de facto barter exchange reminiscent of the preferential trade arrangements of the cold war era. At that time, Cuba received below-market-value oil from the Soviet Union in exchange for highly overvalued Cuban commodities such as sugar, citrus products, rum, and light manufactures. During the cold war, Cuba's primary export was sugar; now it is medical personnel. The recent completion of oil-refining facilities in the southern coast city of Cienfuegos will allow the Venezuelans to export crude oil as opposed to high value refined fuels to Cuba. Another parallel with the Soviet era is that Cuba is again realizing the benefit of excess petroleum products exports, which PDVSA sells to its Petrocaribe customers for hard currency.[35]

In 2008 Brazil's Petrobras was awarded an exploration and production concession for block 37, just north of Cuba's current coastal production area. The project called for an initial budget of $8 million for seismic work during 2009. This marks Petrobras's return from its major Cuban exploration setback in 2001, when its first exploration effort resulted in a dry well.[36] In 2008, a consortium of state- and privately owned Russian oil companies signed a letter of intent with Cupet in order to identify investment opportunities in Cuba's offshore deep waters in the Gulf of Mexico.[37] The financial and technical ability of these Russian companies to invest in and develop Cuba's undiscovered hydrocarbon resources in its northwest EEZ sector seems a bit uncertain at a time when Russia's own domestic oil and natural gas production is declining. Furthermore, the Russian hydrocarbon sector seems to be generally cash-strapped. In 2009, China finalized a $10 billion loan to

Table 2-5. Cuba's Estimated Petroleum Debt to Venezuela, 2003 to 2008

Year	Volume of oil ($ bil)	Weighted acquisition cost[a] ($ bil)	Contractual payment terms (percent)[b]	Ninety-day payment terms (US$ millions)	Twenty-five-year payment terms (US$ millions)	Total (US$ millions)
2003	86.7	28.40	80/20	797	199	996
2004	89.7	40.56	75/25	982	328	1,310
2005	100.5	51.15	70/30	1,363	584	1,947
2006	104.9	53.96	60/40	1,372	915	2,287
2007	98.8	70.24	60/40	1,430	953	2,383
2008	94.0	98.58	50/50	1,691	1,691	3,382
Total				7,635	4,670	12,305

Source: Author compilation, based on data published by Oficina Nacional de Estadísticas de Cuba, 2009.

a. Figures presented are weighted acquisition cost means; the price basis is EIA's U.S. Gulf Coast landed price for Venezuela crude and spot refined product prices.

b. Contractual payment terms and conditions are those outlined in the Convenio Integral de Cooperacion (Convention on Cooperation) between Cuba and Venezuela, signed October 30, 2000, and as amended in July 2008. These are the percentage of profits paid to the joint venture partners. Venezuela's share is represented by the former, Cuba's by the latter.

the Russian oil pipeline monopoly Transneft and another $15 billion loan to the state-run major oil enterprise Rosneft, in exchange for 300 million tons of Russian oil to be delivered over twenty years. Gazprom, Russia's largest company, is also strapped for cash and burdened with a debt of $49.5 billion.[38] There has been questionable management of the Russian oil and gas firms in terms of their ability to sustain positive growth and return on investment in light of the country's considerable oil and gas resources. Russia holds the world's largest natural gas reserves, the second-largest coal reserves, and the eighth-largest oil reserves. Russia is also the world's largest exporter of natural gas, the second-largest oil exporter, and the third-largest energy consumer.

Cuba relied extensively on the largesse of Soviet energy resources for its economic vitality during the cold war, importing nearly 90 percent of all of its oil and gas needs. But Cuba is no longer dependent on Russia to keep its economy afloat—or to plumb the depths of its oil reserves. Repsol and Statoil–Norsk Hydro's commitment in spending an estimated $80 million to $100 million in a new exploratory well, Petrobras's recent entry onto the scene, and, in the background, the U.S. Geological Survey's estimates of undiscovered reserves in the North Belt Thrust—all underscore Cuba's oil and natural gas offshore potential.

Conclusion

It should by now be abundantly clear that Cuba possesses the potential for dramatically increasing its crude oil and refined petroleum production. It stands to reason that if Cuba can successfully extract, refine, and market the 5 billion to 10 billion barrels of oil available in the offshore reserves, it will mark a significant change in the structure of Cuba's energy balance by essentially making it energy self-sufficient. The challenges presented to oil producers operating in deepwater drilling sites in the EEZ are primarily political in nature, in large part because of the obstacle presented by effective American export control regulations regarding the transfer of deepwater drilling technology. Thus, this aspect of the U.S. trade embargo is having a significant impact on Cuba's ability to pursue offshore oil production opportunities. Most if not all of Cuba's partners already possess the technological acumen to drill successfully in deep water, and in some cases these partners are already operating deepwater rigs under contract to U.S. international oil firms in the Gulf of Mexico. As of spring 2010, however, none of those partners is willing to countermand the existing proscriptions against the transfer of this type of technology to Cuba or even its application in Cuban waters.

The economic and political implications of Cuba's becoming not only oil-self-sufficient but also a possible net crude oil and petroleum products exporter could represent a major challenge for future U.S. and Cuban policymakers. The industry's future investment potential—possibly worth tens of billions of dollars—will be determined by the results of exploratory drilling in Cuba's EEZ in the Gulf of Mexico and by U.S. policy toward Cuba.

Appendix Table 1. Crude Oil Liquid Production, Cuba, 2003–09

Cuba—Crude Oil—Liquids Production 2003–2009 (Mt)

	1980	1990	2000	2003	2004	2005	2006	2007	2008	2009	08/09 %
EPEP—Occidente[a]	NA	NA	NA	1,902.2	1,514.4	1,545.8	1,737.7	1,865.0	1,813.2	1,484.6	–18.1
EPEP— Centro[b]	NA	NA	NA	1,735.1	1,701.9	1,354.0	1,129.2	1,009.6	1,161.6	1,219.6	5.0
Natural gas liquids[c]	NA	NA	NA	NA	NA	NA	NA	NA	NA	NA	
Diluents[d]	NA	NA	NA	42.5	36.7	35.3	33.1	30.4	28.3	28.1	
Total liquids production	273.6	670.9	2,695.3	3,679.8	3,253.0	2,935.1	2,900.0	2,905.0	3,003.1	2,731.3	–9.0
Total production (b/d)[e]	4,768	11,672	46,891	64,018	56,593	51,063	50,452	50,539	52,246	47,517	–9.0

Source: Author compilation based on data from Oficina Nacional de Estadísticas de Cuba, 2010.

a. Empresa de Perforacion y Extraccion de Petróleo—Occidente—Cupet's business unit responsible for the exploration and production activities in La Habana and Pinar del Rio provinces, including the production areas of Santa Cruz del Norte, Boca de Jaruco, and Puerto Escondido, among others.

b. Empresa de Perforacion y Extraccion de Petróleo—Centro—Cupet's business unit (165 producing wells) responsible for the exploration and production activities in Matanzas and other central provinces; core activities are located in the Varadero-Cardenas area.

c. ONE does not report condensate or natural gas liquids production. According to Sherritt's financial reports ENERGAS natural gas processing plants located in Varadero and Boca de Jaruco have a combined rated capacity of 70 tpd of sulfur, 684 bpd of condensate, and 913 bpd of natural gas liquids (ethane, propane, butane, and pentane +)

d. Cuban crude oil is too viscous to be easily pumped or transported by pipeline and/or marine vessel. Typically naphtha, condensate, or other diluents are added to improve the viscosity of the crude oil.

e. Barrels per ton conversion factors are determined by the API gravity of the crude oil stream. Cuban crude oil API gravity varies from a low of 8.0–10.0 degrees (Varadero) to a high of 12.0–15.0 (Seboruco, Canasi, Puerto Escondido); we have selected a weighted average conversion factor of 6.35 barrels per ton.

Appendix Table 2. Estimated Market Value of Petroleum Imports, Cuba, 2003–09

	2003			2004			2005			2006			2007			2008			2009 (Est)		
Product	Mbd	$bll	MM$	Mbd	$bll	MM$	Mbd	$bll	MM$	Mbd	$bll	MM$	Mbd	$bll	MM$	Mbd	$bll	MM$	Mbd	$bll	MM$
Crude oil	45.2	25.70	424.0	38.7	33.79	477.3	43.4	47.87	758.3	40.4	57.37	846.0	42.4	66.13	1,034.6	95.5	90.76	3,164.0	96.0	57.70	2,021.8
LPG	2.2	24.19	21.1	3.6	31.15	40.9	3.3	38.39	46.2	2.2	42.58	34.2	1.6	50.81	29.4	2.0	59.37	43.0	2.0	35.46	25.7
Av-gas	0.2	41.87	2.3	0.1	54.39	2.5	0.1	72.37	3.1	0.1	86.35	3.6	0.2	96.81	5.3	0.1	114.00	5.1	0.1	78.79	3.3
Gasoline	0.0	36.59	0.0	1.0	49.13	17.9	0.6	67.04	14.7	1.4	76.68	39.2	3.1	87.66	98.8	4.8	103.79	181.3	5.0	68.68	120.0
Jet-kero	7.2	34.59	90.5	8.5	48.34	150.0	7.4	72.05	194.6	7.0	80.76	206.3	7.2	89.49	235.1	0.0	124.50	0.0	0.0	69.90	0.0
Diesel	17.2	33.77	212.5	19.4	45.46	321.9	20.4	68.24	508.1	22.4	75.94	620.9	22.1	84.32	677.6	12.8	118.06	551.5	13.0	67.94	317.4
Fuel oil	14.9	28.17	153.2	18.5	26.61	179.7	25.4	38.32	355.3	31.5	44.66	513.5	22.4	55.73	453.3	3.0	78.15	85.4	3.0	58.44	64.7
Total	86.9		903.6	89.8		1,190.2	100.6		1,880.3	105.0		2,263.7	99.0		2,534.1	118.2		4,031.4	119.0		2,552.9
WAAC		28.42			40.56			51.15			58.96			69.88			93.41			58.97	
ONE's total			996.3			1,310.4			1,946.5			2,286.6			2,382.8			4,561.8			2,648.7

Source: Author compilation based on data from Oficina Nacional de Estadísticas de Cuba, 2010.

Notes: Price bases are U.S. DOE, Energy Information Administration's U.S. Gulf Coast landed price for Venezuelan crude oil, and for spot domestic refined products.

Volumes are those published by Oficina Nacional de Estadisticas de Cuba (ONE) and PDVSA's financial reports.

WAAC; weighted average acquisition cost. Petroleum imports have a reporting time lag of two years.

Appendix Table 3. Petroleum Demand, Cuba, 2003–09

	Product	2003	2004	2005	2006	2007	2008	2009	09 %V	08/09%
12.00	Refinery gas	1,190	875	1,092	986	588	595	917	0.7	54.1
11.60	LPG	5,234	5,695	6,073	4,135	3,477	3,610	3,645	2.7	1.0
8.90	Aviation gasoline	110	112	100	110	127	97	119	0.1	22.7
8.53	Motor gasoline	8,437	8,310	8,278	8,186	8,287	7,401	6,172	4.5	–19.4
8.22	Naphtha & solvents	709	822	865	1,167	1,180	1,885	2,367	1.7	25.6
	Solvents	NA	NA	NA	NA	NA	NA	NA		
7.93	Jet fuel	3,033	3,726	3,596	3,841	3,641	3,311	2,861	2.1	–13.6
7.73	Kerosene	3,693	5,419	5,360	2,846	2,010	1,838	1,737	1.3	–5.5
7.46	Diesel	24,569	25,626	25,286	29,881	33,057	32,979	29,049	21.2	–11.9
6.66	Industrial fuel oil	98,241	93,955	93,244	91,131	93,282	87,536	87,938	64.1	0.5
	Fuel oil	32,358	34,579	41,641	45,598	41,465	NA	NA		
	Crude blend	65,883	59,376	51,603	45,533	51,817	NA	NA		
7.00	Lubricants	924	959	911	863	779	907	859	0.6	–5.3
6.06	Asphalt	744	745	702	828	903	1,059	1,393	1.0	31.5
5.51	Pet coke	270	196	263	214	116	94	51	0	–45.7
	Total demand	147,154	146,440	145,770	144,188	147,447	141,312	137,108	100	–3.0

Source: Author compilation based on data from Oficina Nacional de Estadísticas de Cuba, 2010.

Appendix Table 4. Refined Products Imports, Cuba, 2003–08

Cuba—Refined Products Imports 2003–2008 (bd)

	Product	2003	2004	2005	2006	2007	2008	07/08 %
11.60	LPG	2,247	3,569	3,337	2,199	1,586	1,983	25.0
8.90	Aviation gasoline	151	127	119	115	151	122	–19.0
8.53	Motor gasoline	0	944	591	1,428	3,106	4,786	54.0
7.93	Jet fuel	7,172	8,523	7,417	7,018	7,239	0	–100.0
7.46	Diesel	17,242	19,445	20,475	22,419	22,137	12,798	–42.0
6.66	Fuel oil	14,895	18,506	25,377	31,521	22,392	3,033	–86.0
7.00	Finish lubes	27	21	31	527	892	662	–26.0
7.00	Base lubes	784	1,015	842	612	426	322	–24.0
	Total imports	42,518	52,150	58,189	65,839	57,929	23,706	–59.0

Source: Author compilation based on data from Oficina Nacional de Estadísticas de Cuba, 2010.

2010: Oficina Nacional de Estadisticas de Cuba

Note: Petroleum imports have a reporting time lag of two years.

Appendix Table 5. Refinery Production, Cuba, 2003–09

Cuba—Refinery Production 2003–2009 (bd)

	Product	2003	2004	2005	2006	2007	2008*	2009*	09 %Y	08/09%
12.00	Refinery gas	1,190	875	1,092	986	861	595	917	0.9	54.1
11.60	LPG	2,946	2,015	2,603	1,970	1,866	1,783	1,468	1.4	–17.7
8.90	Aviation gasoline	0	0	0	0	0	0	0	0.0	0.0
8.53	Motor gasoline	9,635	7,742	9,521	7,415	9,166	16,737	11,498	11.6	–31.3
	Regular	7,565	5,856	6,833	5,361	7,843	13,442	9,661		–28.0
	Premium	2,070	1,886	2,688	2,054	1,323	3,295	1,837		–44.2
8.22	Naphtha	5,353	1,975	2,151	1,889	1,358	1,342	3,770	3.7	181.0
	Solvents	NA	2,840	1,648	2,843	2,574	2,901	2,516	2.4	–13.3
7.93	Jet fuel	0	0	0	198	1,221	6,288	6,414	6.2	2.0
7.73	Kerosene	4,047	4,632	5,604	2,666	1,470	892	500	0.5	–43.9
7.46	Diesel	9,077	7,875	7,466	8,580	9,479	22,411	25,977	25.2	15.9
6.66	Fuel oil	18,690	15,649	15,677	16,280	17,159	48,678	47,976	46.6	–1.4
7.00	Base lubes	33	21	13	15	38	48	38	0.0	–20.8
6.06	Asphalt	744	787	719	828	928	1,023	1,332	1.3	30.2
5.51	Pet coke	270	196	263	214	208	94	51	0.2	–45.7
	Total production	51,985	44,607	46,757	43,884	46,328	102,792	102,457	100.0	0.1
7.00	Finish lubes	919	938	827	867	959	986	719		–27.1

Source: Author compilation based on data from Oficina Nacional de Estadísticas de Cuba, 2010.

2010: Oficina Nacional de Estadisticas de Cuba.

*Note: PDV-Cupet 49%/51% Cienfuegos joint venture 60 mdb refinery on stream.

Appendix Table 6. Petroleum Supply/Demand Balance, 2003–09

Cuba—Petroleum Supply / Demand Balance, 2003–2009 (bd)

		2003	2004	2005	2006	2007	2008	2009	08/09 %
6.35	Crude oil production	64,018	56,593	51,063	50,452	50,539	52,246	47,517	–9.0
	Refinery production	51,985	44,607	46,757	43,884	46,328	102,792	102,457	0.0
7.19	Crude oil imports *	45,208	38,745	43,366	40,429	42,464	95,509	96,000 (Est)	
	Refined products imports	42,518	52,150	58,189	65,839	57,929	23,706	24,000 (Est)	
	Refined products demand	147,154	146,440	145,770	144,188	147,447	141,312	137,108	–3.0
	Inventory built/draw estimate**	4,590	1,048	6,848	11,476	2,398	30,149	30,409 (Est)	
	Reported exports***	NA	–232	NA	–1,740	–1,456	–622	NA	
6.29	Natural gas production (boed)	11,339	12,135	12,808	18,794	20,988	20,007	20,099	0.5

Source: Author compilation based on data from Oficina Nacional de Estadísticas de Cuba, 2010.

2010: Oficina Nacional de Estadisticas de Cuba.

*Crude oil imports conversion factor used is 7.19 barrels per ton based on Mesa 30 Venezuelan crude oil blend.

**Cuba does not report petroleum inventory levels. Primary, secondary, and tertiary working inventory levels for crude oil and petroleum products should be around 15–30 days of anticipated demand. Any inventories over the working level should be deemed to be exports.

***Cuba does not report petroleum products exports. These volumes are third countries' reported Cuban petroleum products imports.

Note: Petroleum imports have a reporting time lag of two years.

Appendix Table 7. Cuba-Venezuela Estimated Petroleum Debt, 2003–08

2003–2008 Cuba-Venezuela Estimated Petroleum Debt ($mmUSd)

Year	Volume	WAAC	Terms	90-day	25-year	Total
2003	86.7 mbd	$28.40	80/20%	$797	$199	$996
2004	89.7 mbd	$40.56	75/25%	$982	$328	$1,310
2005	100.5 mbd	$51.15	70/30%	$1,363	$584	$1,947
2006	104.9 mbd	$53.96	60/40%	$1,372	$915	$2,287
2007	98.8 mbd	$70.24	60/40%	$1,430	$953	$2,383
2008	99.5 mbd	$93.41	50/50%	$1,696	$1,696	$3,392
2009	92.5 mbd	$58.97	60/40%	$1,195	$796	$1,991
Total				**$8,835**	**$5,471**	**$14,306**

Source: Author compilation based on data from Oficina Nacional de Estadísticas de Cuba, 2010.

WAAC; weighted acquisition cost; price bases are EIA's US. Gulf Coast landed price for Venezuela crude and spot refined product prices. Contractual payment terns and conditions are those outlined in the *Convenio Integral de Cooperacion* between Cuba and Venezuela signed October 30, 2000, and as amended July 2008.

Notes

1. Jose Olivares, *Our Islands and Their People* (New York: N. D. Thompson, 1899).

2. Alvaro Franco, "Que Hay en Cuba?" [What is there in Cuba?] *Petróleo Internacional,* July–August, 1993.

3. "First License Round in Thirty Years Sheds Light on Cuba's Geology," *Oil and Gas Journal,* April 26, 1993.

4. The production-sharing agreement (PSA) offers an alternative to the joint venture as a way for two or more economic entities to collaborate on the development and production of a commodity. The PSA is the principal way for foreign firms to invest in a country while the country retains a degree of control over valuable resources. Under a standard form of PSA, the entity that invests in a development project is the first to capture the investment from revenues generated by the forthcoming output.

5. Ibid.

6. Jon Frederic Blickwede, "Cuba: An Overview of Its Geology, Hydrocarbon System, and Petroleum Industry," paper presented at the American Association of Petroleum Geologists, 2002 Annual Meeting, Houston (March 10–12, 2002).

7. Manuel Marrero Faz, "Estado Actual y Perspectivas de la Exploración y Producción de Petróleo y Gas en Cuba" [Current status and prospects for the exploration and production of oil and natural gas in Cuba], Second Cuban Congress on Oil and Gas, Havana (March 17, 2009).

8. The American Petroleum Institute has established a standard measure, called API gravity, for the density of oil. The American Petroleum Institute gravity, or *API gravity,* is a measure of how heavy or light a petroleum liquid is compared to water. If a substance's API gravity is greater than 10, it is lighter and floats on water; if less

than 10, it is heavier and sinks. For further information see http://widman.biz/English/Calculators/DegreesAPI.html.

9. The abbreviation EBITDA is an indicator of a company's financial performance. See Sherritt, Inc., *2008 Financial Annual Report* (Montreal: 2009).

10. Pebercan Inc., *First Quarter 2009 Financial Report* (Montreal: 2009).

11. Sherritt, Inc., *Second Quarter 2008 Financial Report* (Montreal: 2008).

12. U.S. Department of Energy, "Undeveloped Domestic Oil Resources: The Foundation for Increasing Oil Production and a Viable Domestic Oil Industry" (Washington, D.C.: February 2006).

13. *Granma Internacional,* February 21, 2005. See also Ria Novosti, "Rusia Invertirá 6,5 Milliones de Euros en la Explotación de Cuatro Bloques Petroliferos en Cuba," *La Habana,* December 11, 2009.

14. "Petro China's Great Wall Drilling Co. Wins Contract in Cuba," *China Chemical Report,* November 26, 2005.

15. Erika Bolstad, "GOP Claim about Chinese Oil Drilling off Cuba Is Untrue," McClatchy Newspapers, June 11, 2008.

16. Yusneurys Pérez Martínez, "Análisis del Potencial Petrolero en el Sector Alamar-Bacuranao" [Analysis of the oil potential in the Alamar-Bacuranao sector], paper presented at the Second Cuban Congress for Oil and Gas, Havana (March 17, 2009).

17. Oficina Nacional de Estadísticas de Cuba, *Anuario Estadístico de Cuba 2007–2007* (Havana: ONEC, 2007).

18. Sherritt, Inc., *2008 Financial Annual Report.*

19. No pre-2007 datafiles exist.

20. Cámara Petrolera de Venezuela, "Proyectos a Desarrollar en Cuba" [Development projects in Cuba], Press Release, Caracas, October 21, 2008.

21. Guillermo Hernandez Perez, "Cuba Deepwater Exploration Opportunities Described in Southeastern Gulf of Mexico," *Oil and Gas Journal,* December 11, 2000.

22. U.S. Department of Interior, Mineral Management Services, "Report to Congress: Comprehensive Inventory of U.S. OCS Oil and Gas Resources: Energy Policy Act of 2005—Section 357" (Washington: February 2006) (www.mms.gov/revaldiv/PDFs/FinalInventoryReportDeliveredToCongress-corrected3-6-06.pdf).

23. U.S. Department of State, Office of Ocean Law and Policy, Bureau of Oceans and International Environmental and Scientific Affairs, "Maritime Boundary: Cuba–United States," paper no. 110, Limits in the Seas (Washington: February 21, 1990) (www.law.fsu.edu/library/collection/LimitsinSeas/ls110.pdf).

24. U.S. Department of the Interior, U.S. Geological Survey, "Assessment of Undiscovered Oil and Gas Resources of the North Cuba Basin, Cuba, 2004," World Assessment of Oil and Gas Fact Sheet (Washington: February 2005) (http://walrus.wr.usgs.gov/infobank/programs/html/factsheets/pdfs/2005_3009.pdf).

25. The North Cuba Fold and Thrust Belt AU was defined to encompass all geological structures within the Late Cretaceous-Paleogene fold and thrust belt in northwestern Cuba. The North Cuba Foreland Basin AU includes all potential oil

and gas accumulations in the foreland basin that formed in front of the thrust belt, and it includes oil and gas accumulation in potential rift-related structures below the rocks of the foreland basin. See U.S. Department of Energy, Assessment of Undiscovered Oil and Gas Resources of the North Cuba Basin, February 2005 (http://pubs.usgs.gov/fs/2005/3009/pdf/fs2005_3009.pdf).

26. Manuel Marrero Faz, "Estado Actual y Perspectivas de la Exploración y Producción de Petróleo y Gas en Cuba" [Current status and perspectives on Oil and Gas Exploration and Production in Cuba], paper presented at the Second Cuban Congress on Oil and Gas, Havana (March 17, 2009).

27. "Repsol confirma perforación de pozo petrolero en Cuba" [Repsol confirms it is drilling an oil well in Cuba], *América Económica,* July 29, 2004.

28. "Repsol lo intenta por segunda vez en Cuba" [Repsol tries for the second time in Cuba], *Expansíon,* June 3, 2005.

29. Jeff Franks, "Repsol Has Contract for Oil Rig, Said Cuba-Bound," Reuters, May 5, 2010.

30. "Repsol Makes New Oil Find in Gulf of Mexico," *Alexander's Gas & Oil Connections,* March 5, 2009.

31. Ron Scherer, "Russian Oil Rigs Just 45 Miles from Florida?" *Christian Science Monitor,* August 5, 2009.

32. A related analysis can be found in Jorge R. Piñón-Cervera, "Cuba's Energy Challenge: A Second Look," *Cuba in Transition* 15 (2005): 110–23.

33. "Venezuela Pays USD 5.6 billion for Cuban Staff in 2008," *El Universal* (Caracas), October 5, 2009.

34. Convenio Interal de Cooperacion Cuba-Venezuela, Caracas, October 30, 2000.

35. Mark Frank, "Oil Now Second-Leading Cuba Export–Gov't Report," Reuters, June 10, 2009.

36. "Petrobras May Commence Drilling Offshore Cuba within a Year," *Alexander's Gas & Oil Connections,* March 19, 2009 (www.gasandoil.com).

37. Rodrigo Fernández, "Raúl Castro abre una nueva página en las relaciones entre Rusia y Cuba" [Raúl Castro opens a new page in Russia-Cuba relations], *El País,* Moscow, January 30, 2009.

38. Andrew E. Kramer, "Russia's Once Mighty Oil Giant Gazprom Buried in Debt," *New York Times,* December 29, 2008.

three

The Electric Power Sector in Cuba: Ways to Increase Efficiency and Sustainability

JUAN A. B. BELT

Since the beginning of the Cuban Revolution, the electric power sector of Cuba has been managed with little regard for financial and economic issues. This approach to the power sector has a long history in the Communist bloc, as the sector was considered to have a preeminent political dimension. Lenin famously noted in 1920, "Communism is Soviet power plus the electrification of the whole country."[1] It was not considered necessary nor even desirable for this sector to pay for itself or turn a profit.

Cuba's per capita annual power consumption is about 1,300 kilowatt-hours (kWh). Table 3-1 shows Cuba's per capita GDP and per capita electricity consumption in comparison with three other countries.[2] (This

The author would like to thank two of his colleagues at the U.S. Agency for International Development (USAID), Micah Globerson and Luis Velazquez, for their great support in the preparation of this chapter, and Silvia Alvarado, Ignacio Rodriguez, and Elizabeth P. Belt for providing comments. The author would also like to thank the International Resources Group team of Evelyn Wright, Gary Goldstein, Pat Delaquil, and Adam Chambers, who modeled the power sector of Cuba using the MARKAL/TIMES platform. Working with the IRG team was a learning experience—and a pleasure.

The opinions expressed in this paper are those of the author and do not represent the views of the U.S. government. A number of the ideas discussed in this chapter were presented at different annual meetings of the Association for the Study of the Cuban Economy. The author bears sole responsibility for any errors in this chapter.

Table 3-1. Per Capita Gross Domestic Product and Electricity Consumption—Comparison of Chile, Costa Rica, Cuba, and Dominican Republic, 2008

	Gross domestic product		Electricity consumption	
Country	PPP (U.S.$)[a]	Nominal (U.S.$)[b]	Percentage coverage	Consumption (kilowatt-hours)
Chile	14,552	10,768	99	3,421
Costa Rica	10,768	6,563	99	1,786
Cuba	9,680	4,924	95	1,214
Dominican Republic	8,222	4,619	92	1,297

Source: CIA, *The World Factbook*, 2009. There is significant controversy on Cuba's National Income Accounts and different publications show very different figures for GDP. As the CIA *Factbook* is one of the few publications with data for the four countries, these data were used to construct the table.

a. GDP at Purchasing Power Parity.

b. GDP converted to U.S. dollars using the nominal exchange rate.

chapter presents electric power data comparing Chile, Costa Rica, Cuba, and the Dominican Republic.)

The power sector in Cuba is controlled almost entirely by the state; the only private participation is in the form of independent power producer arrangements. The situation is somewhat different in the hydrocarbon sector, as there has been significant private participation in exploration and crude oil production since the 1990s. This is the result of the very attractive production-sharing agreements (PSAs) offered in Cuba. But international trade in oil and derivatives—refining, distribution, and pricing—do not seem to follow financial and economic considerations there. This lack of concern for financial and economic pressures may result from the fact that both the Soviet Union and Venezuela have provided large subsidies to Cuba by supplying oil and derivatives on concessionary terms, and Cuba's economic policymakers do not seem to take account of opportunity costs in the energy sector (see also chapter 2 of this volume).

The power and hydrocarbon sectors are inextricably linked, as Cuba produces about 85 percent of its power using liquid fuels, a very high percentage compared with other countries.[3] The total value of the energy consumed in Cuba has been estimated at 14 percent of GDP, compared with a world average of about 10 percent. In 2007, domestic production of crude oil accounted for about 40 percent of total consumption and the rest was imported from Venezuela. About 50 percent of the total supply of fuel oil is applied to power generation and 50 percent for transportation and other uses; this is consistent with the usage breakdown seen in other countries.

Table 3-2. Cuba's Liquid Fuel Supply, 2007

Liquid fuel sources	Billions of barrels/day	Percentage of total
Domestic production	68,000	40
Imports	102,000	60
Total supply	170,000	100
Liquid fuel uses		
Power generation	85,700	50
Transport	84,300	50
Total uses	170,000	100

Source: Author's compilation based on Manuel Cereijo, "Cuba's Power Sector: 1998 to 2008," paper presented at the 18th annual meetings of the Association for the Study of the Cuban Economy, Miami (August 2008), and Oficina Nacional de Estadísticas de Cuba, *Anuario Estadístico de Cuba 2008* (Havana: 2009), tables 10.1 through 10.19.

Main Trends in the Cuban Power Sector

Until 1959, Cuba had four electric utilities, all regulated by the Public Service Commission under the Ministry of Communications, with additional power supplied by the large sugar mill industry. Most of the boilers and turbine generators used to produce energy came from the United States or West Germany. After the revolution, the entire sector (generation, transmission, and distribution) was nationalized and absorbed into a state-owned utility, Unión Eléctrica, which is under the Ministry of Basic Industry (Minbas).

The development of the sector since the revolution can be divided into three distinct periods:[4]

—1959 to 1989. These three decades, beginning with Castro's takeover and ending with the fall of the Soviet Union, saw rapid growth in Cuba's energy sector, facilitated by subsidized Soviet oil imports and other forms of financial support. The period included the country's largest buildup in energy generation infrastructure and the highest rates of growth in consumption, made possible by oil and products imported from the Soviet Union at highly subsidized prices.

—1990 to 1997. This was the so-called special period (*periodo especial*), after the breakup of the Soviet Union, when the Cuban economy was subjected to extreme survival pressure with sharp declines in GDP and the introduction of some measures to promote private sector activity. Domestic oil production accelerated and Cuba began to use fuel oil in the seven large generation plants. Unfortunately, the domestic oil's high sulfur levels severely damaged the generation infrastructure. The special period came to an end when the economy stabilized, partly as a result of massive Venezuelan financial support.

—1998 to 2010. This most recent period has been characterized by Venezuelan support, the blackouts of 2004–05 (caused by power plant breakdowns), the energy revolution (*revolución energética*) of 2005–06 (a program to reduce electricity consumption and to expand generation capacity), and the independent power production arrangement with a Canadian company, Sherritt, whereby power is being produced by combined-cycle gas turbines supplied by that company. The *revolución energética* was successful in introducing energy efficiency, but the increases in generation, with the exception of the Sherritt deal, are based on small generator sets, called gensets, that are highly inefficient. The sector is significantly more stable now than it was during the period of the blackouts. Still, the high proportion of generation from burning liquid fuels results in extremely high costs and very high carbon emissions. The financial sustainability of the sector depends almost totally on the largesse of Venezuela. If support from Venezuela were reduced or terminated, the power sector would require extremely high subsidies. The government of Cuba does not have adequate fiscal resources for this kind of subsidy, and it would be forced to embark on an economic reform effort more comprehensive than the one experienced during the *periodo especial,* following the dissolution of the Soviet Union.

These trends will be discussed in greater detail, with particular emphasis on the most recent period.

Installed power generation capacity increased from less than 400 megawatts (MW) in 1958 to about 4,000 MW in 1990, an annual compound rate of growth of almost 12 percent. During the same period, total electricity consumption grew from about 1,500 gigawatt-hours (GWh) to about 9,700 GWh, an annual growth rate of 6 percent (see figure 3-1).

When Cuba lost Soviet assistance, which was estimated at $5 billion to $7 billion annually, it suffered a sharp decline in GDP, accompanied by a rapid decline in energy consumption per capita in the period from 1990 to 1995 (see figure 3-2). A drop in consumption by "industry and construction" accounted for most of the decline, as other types of consumers, including households, did not curtail consumption.

By 2005 and 2006, power plant breakdowns caused a wave of severe blackouts, sometimes lasting up to eighteen hours per day, which led to civil unrest. As a result of the blackouts the government of Cuba embarked on the *revolución energética,* to reduce electricity consumption and to expand generation capacity.

Figure 3-1. Cuba's Installed Generation Capacity

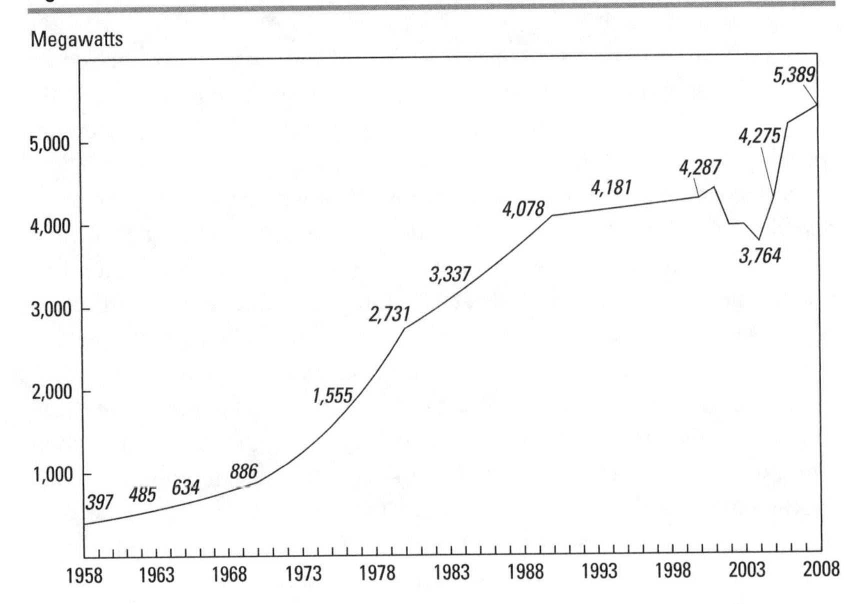

Source: Author's compilation based on data from Oficina Nacional de Estadísticas de Cuba, *Anuario Estadístico de Cuba 2006* (Havana: 2006), chapter 8, "Energía" (www.one.cu/aec_web/paginas_de_textos/c_viii.htm).

Figure 3-2. Electricity Consumption, by Key Sector, 1958–2008

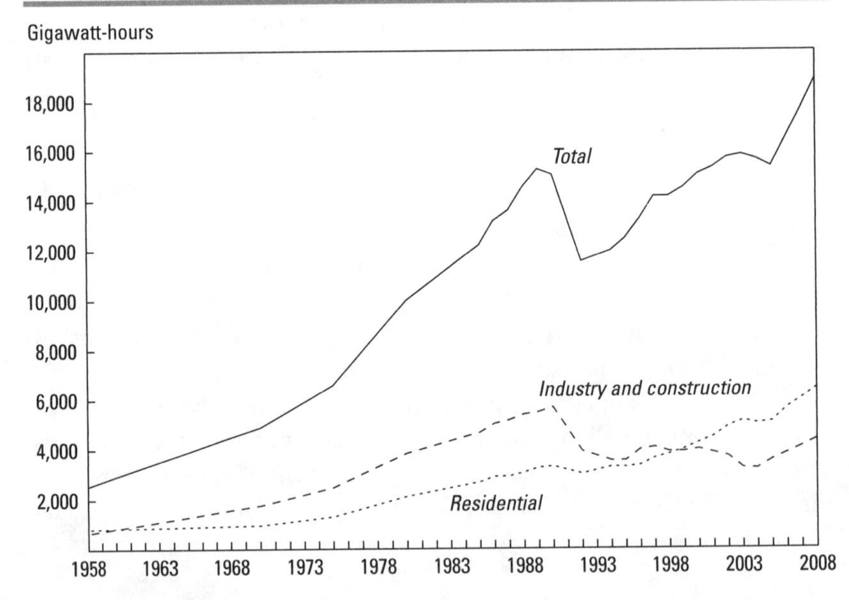

Source: Author's compilation, based on data from Oficina National de Estdísticas de Cuba, *Anuario Estadístico de Cuba 2006.*

The *revolución energética,* which largely emphasized energy conservation, was reasonably successful in reducing daily peak demand. The conservation component of this program involved removing individuals' energy-inefficient refrigerators, fans, and air conditioners, and replacing them with energy-efficient appliances and light bulbs. These upgrades were somewhat compulsory, and were incentivized in various ways, including the use of payment plans of up to ten years for the purchase of improved devices, with payments discounted directly from salaries.[5]

The results of the initiatives to expand capacity, on the other hand, are mixed. The two principal elements of the production component of the energy revolution were the installation of a new distributed generation system ("distributed generation" refers to generating power close to the point of consumption) based on small gensets, called *grupos electrógenos,* that use thousands of small fuel-burning generators, and the erection of gas-fired generation plants established under a power purchase agreement (PPA) with Sherritt. The latter measure has been a highly positive development that has significantly reduced both generation costs and carbon emissions.

But the distributed generation project had distinct drawbacks. Distributed generation is an extremely costly source of power if fuel is valued at its international price. The distributed approach employed a couple thousand small gensets scattered around nearly 70 percent of the island's 169 municipalities. The mini-generators had been produced in South Korea, Germany, and Spain.[6] This investment began in 2005 and continues to provide a high-cost, low-efficiency 1.5 GW of additional energy output. Cuba also established a small wind power plant that produces about 20 megawatts; Cuba is reportedly also considering other renewable options, such as ethanol fuel from sugarcane.

The *grupos electrógenos* approach has special benefits for Cuba but also some significant costs. Distributed generation offers increased production flexibility in the face of hurricanes, which have been particularly frequent in recent years, averaging more than one per year over a fifteen-year period. Whereas the shutdown of a large generator would have an immediate impact on a delicate energy system, individual failures of smaller generators will have a more localized effect in a smaller region. Thus, the increased number of individual sources reduced overall systemic risk of a large blackout. The smaller units were also simple to install and put into service quickly, immediately ending blackouts and muting public disquiet. There are, however, distinct disadvantages to this approach: very high operating costs, at the normal international price of diesel oil and fuel to run the generators; the challenge of providing maintenance and service to thousands of generator

Figure 3-3. Per Capita Electricity Consumption—Comparing Cuba, Chile, Costa Rica, and Dominican Republic

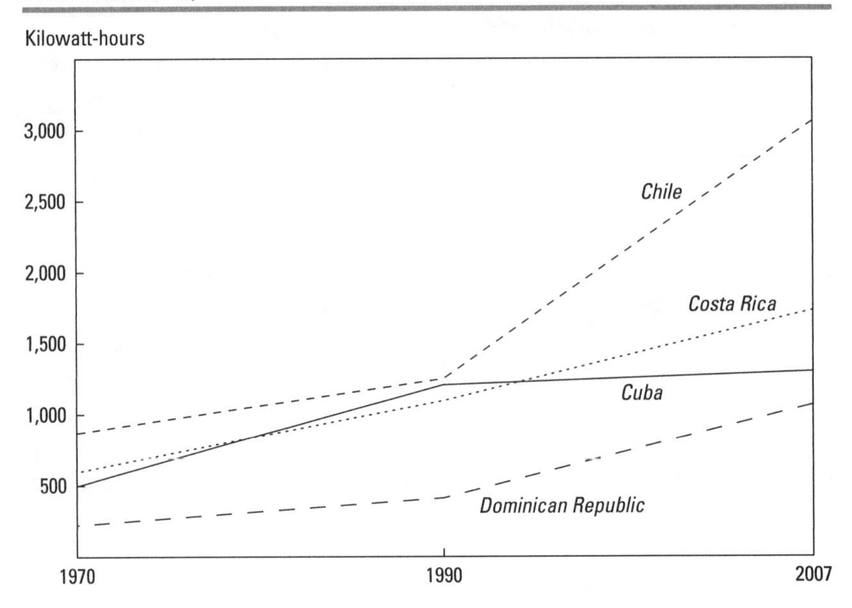

Source: Author's compilation, based on Central Intelligence Agency, *The World Factbook* (www.cia.gov/library/publications/the-world-factbook).

units; the difficulty of efficiently dispatching numerous scattered generators (in a power system, there is always a dispatch center, which orders different generators to provide electricity to the grid); and the probability that transmission stability problems will arise.

It is interesting to compare per capita electricity consumption rates in different countries (see figure 3-3). Note that whereas Costa Rica, Chile, and Cuba had similar consumption per capita in 1990, consumption in Chile and Costa Rica today is significantly higher than in Cuba. Chile, which relies mostly on private investment in the power sector, increased generation much faster than Costa Rica, where the private sector plays only a minor role in the power sector.

Financial and Economic Aspects of Unión Eléctrica

Unión Eléctrica, the vertically integrated utility providing power to most of the country, was established in 1960. Financial data for Unión Eléctrica are not available, so the financial analysis presented in this section should be considered approximate.[7] More research is needed to refine the numbers—an important task, given the centrality of the energy sector to the Cuban economy in general and to the fiscal accounts in particular. A further challenge in understanding Unión Eléctrica's financial profile is the dual monetary system—the use of both pesos and dollars in the Cuban economy. This imposes important conceptual difficulties, as it is very difficult to compare financial flows in U.S. dollars with financial flows in pesos. We use the official exchange rate of one Cuban peso (1 CUP) to one dollar, but we carry out sensitivity analysis to determine the effects on our conclusions of different exchange rates. Furthermore, as most workers in Cuba are employed by the state or by state-owned enterprises, the government essentially establishes wages throughout the economy, and consequently it is very difficult to speculate about the opportunity cost of labor—what the labor would have produced in its best alternative use.[8]

The following analysis, developed from multiple sources, is a rough approximation of Unión Eléctrica's actual financial situation. The main parameters and assumptions used to estimate the 2008 cash flow are the following (all dollar amounts are U.S. dollars unless otherwise indicated):

—Total value of sales: $2.8 billion.[9] Applying present rates to net sales (total sales minus losses and minus plant load) results in a figure of $2 billion. The higher figure, which came from the analysis of Manuel Cereijo, was used in this analysis; if the lower figure were used, the results would be much more negative.

—Number of workers: 33,950

—Total wage bill: CUP 266 million

—Total use of fuel for power generation: 28 million barrels

—Average cost of crude oil at international prices in 2008: $100 per barrel

—Average cost of power generation fuel at international prices in 2008: $125 per barrel

—Estimated total value of assets: $6.8 billion

—Annualized capital cost, at 10 percent interest, for twenty years: $794 million

—Operations and maintenance costs: 1.5 percent of total asset value: $101 million

In 2008, the economic profit of Unión Eléctrica was estimated to be negative, a loss of $1.9 billion, with an EBITDA (earnings before interest, taxes, depreciation, and amortization) of approximately –$1.1 billion. If the value of Venezuelan fuel is discounted by 40 percent, EBITDA becomes

Table 3-3. Oil Price Forecasts for 2009, as of January 15, 2009

Morgan Stanley	$82	CIBC World Markets	$50
Barclays Bank	$76	Goldman Sachs Group	$45
Analyst consensus (Reuters)	$58	JP Morgan Securities	$43
U.S. Dept. of Energy	$51	Purchasingdata.com	$43

Source: "Oil Price Forecasts Stay Low" (www.purchasing.com/article/CA6628617.html).

positive, around $329 million.[10] Unión Eléctrica's situation improved following the sharp decline in oil prices in late 2008. At current crude oil prices of about $70 per barrel and assuming a constant refining margin of $25 per barrel, an even exchange rate of CUP 1 = $1, and a 40 percent fuel discount, EBITDA was estimated at a positive $833 million. Many allege that Cuba pays nothing for Venezuelan oil. If that is the case, the EBITDA for Unión Eléctrica may exceed $2.4 billion, even if Cuba pays international prices for the domestic oil that it purchases through private companies participating in the production-sharing agreements.

Table 3-3 shows 2009 crude oil price forecasts from eight major institutions, ranging from $43 to $82 per barrel. If the average price of crude oil reaches $70 per barrel this year, which is roughly the average of the estimates, Unión Eléctrica faces economic losses of $1 billion (see table 3-4),

Table 3-4. Unión Eléctrica's Estimated Financial Results (Variation in Profit and EBITDA as a Function of the Price of Crude Oil), 2008[a]

	Price of crude oil per barrel ($)				
	30[b]	40[c]	60	87[d]	100[e]
Fuel price (80 percent fuel oil, 20 percent diesel) ($/barrel)	55	65	85	112	125
Economic profit ($ millions)	97	–184	–746	–1,504	–1,870
EBITDA (fuel at international price) ($ millions)	891	610	48	–710	–1,076
EBITDA (fuel discounted 40 percent) ($ millions)	1,508	1,340	1,003	548	329
EBITDA (fuel discounted 100 percent) ($ millions)	2,436	2,436	2,436	2,436	2,436

Source: Author's compilation based on Manuel Cereijo, "Cuba's Power Sector: 1998 to 2008," and Oficina Nacional de Estadísticas de Cuba, *Anuario Estadístico de Cuba 2008.*

a. This table assumes a uniform refining margin of $25 per barrel of oil in 2008.

b. Crude oil price for Unión Eléctrica's breakeven point is estimated at approximately $33.30/barrel.

c. Average crude oil price in 2009 (two months only).

d. Average crude oil price in 2007.

e. Average crude oil price in 2008.

Figure 3-4. Economic Profit as a Function of Crude Oil Price, 2008

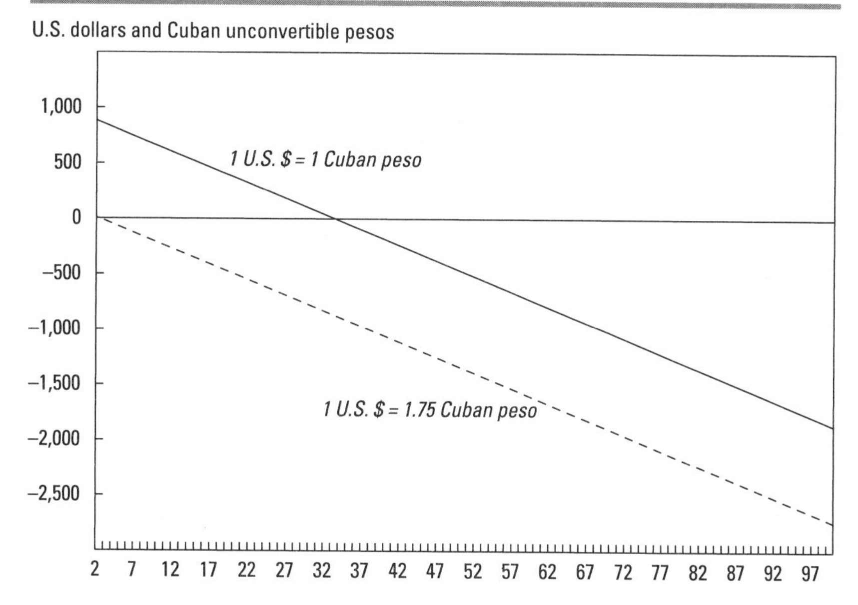

Source: Author's compilation, based on data from Oficina Nacional de Estadísticas de Cuba, *Anuario Estadístico de Cuba 2006.*

and EBITDA would be a negative $233 million, if Cuba were to pay the full international price.

Unión Eléctrica's results depend critically on the market price of crude oil. For most countries that, like Cuba, base their power generation largely on liquid fuels, a reduction of the price would be good news. In the case of Cuba, however, a significantly lower price for crude oil may also lead to energy insecurity if low oil prices also result in a major reduction in Venezuelan subsidies. If the crude oil price declines, Venezuela's dire economic situation would deteriorate further, making it harder to continue to subsidize Cuba's energy needs. This highlights a critical vulnerability in the Cuban power sector: its dependence upon a stable relationship with its main foreign donor, Venezuela.

Economic profit and EBITDA were calculated over a range representing historical crude oil prices in 2007 and 2008, as well as prices in 2009 (see table 3-4). Oil price forecasts fall throughout this range, but average around $56 per barrel. Unfortunately, these calculations show Cuba as having an economic profit shortfall as long as prices exceed $33, which is an amount below the current international market price for crude. The effect of oil prices on economic profit is shown in figure 3-4, which also depicts the relationship

Table 3-5. Economic Profit as a Function of Crude Oil Prices and Exchange Rates: Sensitivity Analysis for Exchange Rates, Crude Oil, and Fuel Prices

		Exchange rates (CUP/U.S. $1)			
		1	1.5	1.75	2
Unión Eléctrica sales revenue and economic profit (million U.S.$)		2,803	2,015	1,790	1,621
Net economic profit (π) when crude oil price ($/bbl) is . . .	and fuel price, after refining (80% fuel oil, 20% diesel) ($bbl), is				
2	27	883	204	10	–136
5	30	799	120	–74	–220
10	35	658	–21	–215	–360
15	40	518	–161	–355	–501
20	45	378	–302	–496	–641
25	50	237	–442	–636	–782
30	55	97	–583	–777	–922
40	65	–184	–864	–1,058	–1,203
60	85	–746	–1,425	–1,619	–1,765
87	112	–1,504	–2,184	–2,378	–2,523
100	125	–1,870	–2,549	–2,743	–2,889

Source: Author's compilation based on Manuel Cereijo, "Cuba's Power Sector: 1998 to 2008," and Oficina Nacional de Estadísticas de Cuba, *Anuario Estadístico de Cuba 2008.*

between the economic profit and the price of crude oil under the assumption that the equilibrium exchange rate is 1.75 CUP = $1.0. At this rate of exchange, the break-even price for crude oil is slightly above $2.

The data that underlie the graph in table 3-4 are presented in table 3-5, which estimates economic profit as a function of the fuel price and of the exchange rate. If the "correct" exchange rate is 1.75 CUP = $1.0 and if the price of oil were to be US$70 per barrel (average forecast of seven institutions is presented in table 3-3), the economic losses of Unión Eléctrica would be about $1.0 billion.

Unión Eléctrica is an inefficient enterprise by international standards, with very low labor productivity, high technical losses, and an overreliance on liquid fuels. Table 3-6 summarizes key indicators for Unión Eléctrica and compares the Cuban power sector to those of Chile, Costa Rica, and the Dominican Republic. The ratio of employees to 1,000 connections at Unión Eléctrica is 9.0, compared to 5.1 in the Dominican Republic, 3.8 in Costa Rica, and 0.7 in Chile. Losses (mostly technical, as there is almost no energy theft in Cuba) are 143 percent of those in Chile and 88 percent of those

Table 3-6. Productivity and Efficiency of Power Sectors—Comparison of Chile, Costa Rica, Dominican Republic, and Cuba

Indicator	Chile	Costa Rica	Dominican Republic	Cuba
Total number of connections	4,861,913	1,236,847	914,279	3,923,650
Total number of residential connections	4,486,053	1,080,591	844,613	3,773,720
Total electricity sold per year (megawatt-hours)	29,000,000	11,800,000	3,719,640	13,892,760
Electricity sold per connection (megawatt-hours/year)	6.5	10.9	4.4	3.7
Total electricity losses (percent)	6.5	8.4	42.5	15.8
Total employees in power sector	3,136	4,155	4,317	33,949
Employees per 1,000 residential customers	0.7	3.8	5.1	9

Source: World Bank, "Benchmarking Data of the Electricity Distribution Sector in the Latin American and Caribbean Region, 1995 to 2005" (http://info.worldbank.org/etools/lacelectricity/home.htm).

in Costa Rica. Losses in the Dominican Republic, a country with massive energy theft, are much higher than in Cuba.[11] In 2003, Cuba ranked fifth in the world for percentage of total energy derived from liquid fuels. This excessive dependence results in very high unit generation costs, as fuel accounts for 70 to 80 percent of total generation costs. Table 3-7 shows unit electricity costs as a function of crude oil prices. Even if losses were eliminated, costs per kWh would be about $0.33 (with crude oil priced at $100 per barrel)—a cost level that would make just about any economic activity uncompetitive in international markets.

At present crude oil prices, fuel accounts for more than 60 percent of total Cuban electricity costs (see table 3-8). Increases in labor productivity and

Table 3-7. Unit Costs of Electricity—Sensitivity Analysis to Crude Oil Prices, 2008[a]

Crude oil price ($/bbl)	Fuel price (80/20 percent fuel oil/diesel) ($/bbl)	Cost per kilowatt-hour sold ($/kWh)	Cost per kilowatt-hour generated ($/kWh)
30	55	0.1943	0.1532
40	65	0.2145	0.1691
60	85	0.2549	0.2009
87	112	0.3093	0.2439
100	125	0.3356	0.2646

Source: Author's compilation based on Manuel Cereijo, "Cuba's Power Sector: 1998 to 2008," and Oficina Nacional de Estadísticas de Cuba, *Anuario Estadístico de Cuba 2008*.

a. This table assumes a uniform refining margin of $25 per barrel of oil in 2008.

Table 3-8. Percentage Production Costs as a Function of the Price of Crude Oil, 2008
Percent of price per barrel

	Price of crude oil per barrel ($)			
Cost	30	40	60	87
Labor	9.80	8.90	7.50	6.20
Fuel	57.10	61.10	67.30	73.00
Operations and maintenance	3.70	3.40	2.90	2.40
Annualized capital	29.30	26.60	22.40	18.40
Total costs (fuel 100%)	100.00	100.00	100.00	100.00

Source: Author's compilation based on Manuel Cereijo, "Cuba's Power Sector: 1998 to 2008," and Oficina Nacional de Estadísticas de Cuba, *Anuario Estadístico de Cuba 2008.*

reductions in losses can reduce costs per kWh, but major decreases are not possible unless fuel costs decline through greater plant efficiency and by moving to other less costly sources of power generation, including renewables. Tables 3-9 and 3-10 summarize the sensitivity analysis that shows the effect of different exchange rates on Unión Eléctrica's financial and economic results. Not surprisingly, higher exchange rates make the situation worse.

Future Development Prospects for the Cuban Power Sector

The Cuban power sector will require extensive investment, competent and professional management, and sector restructuring to reduce high operat-

Table 3-9. Economic Profit as a Function of the Exchange Rate[a]

	Sales revenue		Totals in U.S.$ under varying exchange rates (CUP/$1)			
Revenue and expenditures	Local (CUP)	Foreign (U.S.$)	1	1.5	2	3
Sales revenue	2,365	438	2,803	2,015	1,621	1,227
Expenditures						
Labor	266	0	266	177	133	89
Fuel	0	3,511	3,511	3,511	3,511	3,511
Annualized capital		794	794	794	794	794
Operations and maintenance	61	41	101	81	71	61
Total expenditures	327	4,346	4,673	4,564	4,509	4,455
Economic profit	2,038	−3,908	−1,870	−2,549	−2,889	−3,228
Cash flow			−1,076	−1,755	−2,095	−2,434

Source: Author's compilation based on Manuel Cereijo, "Cuba's Power Sector: 1998 to 2008,"and Oficina Nacional de Estadísticas de Cuba, *Anuario Estadístico de Cuba 2008.*

a. Calculated for average crude oil price of $100/barrel and a refining margin of $25/barrel.

Table 3-10. Breakeven Rates for Equilibrium as a Function of Exchange Rates[a]

	Rate of exchange (CUP/$1)			
	1	1.5	2	3
Total sales revenue (U.S.$ millions)	2,803	2,015	1,621	1,227
Net energy sold (gigawatt-hours)	13,925	13,925	13,925	13,925
Total costs (U.S.$ millions)	3,549	3,440	3,386	3,331
Unit sales revenue (U.S.$/kilowatt-hour)	0.201	0.145	0.116	0.088
Breakeven rate (U.S.$/kilowatt-hour)	0.255	0.247	0.243	0.239
Economic profit (U.S.$ millions)	0	0	0	0

Source: Author's compilation based on Manuel Cereijo, "Cuba's Power Sector: 1998 to 2008," and Oficina Nacional de Estadísticas de Cuba, *Anuario Estadístico de Cuba 2008.*

a. This table assumes a uniform refining margin of $25 per barrel of oil in 2008.

ing costs and to meet growing demand in the coming years.[12] As discussed in previous sections, power generation in Cuba relies primarily on a set of aging oil-burning plants, whose condition has been compromised by the burning of heavy, sour domestic crude oil. These plants have been supplemented recently by natural gas plants, financed through international joint ventures and by new, small gensets generating a total of more than one gigawatt (GW). Going forward, the system is subject to a large number of uncertainties, including the rate of economic growth and international fuel prices. Other uncertainties—such as the rate and nature of market liberalization, openness to and availability of foreign investment, and changes in the structure of energy demand—are particular to the Cuban situation. Decisions about investment in new generation capacity will involve weighing several interlocking factors: capital versus operating costs, future fuel cost risk, future demand growth, demand-side versus supply-side investment, availability of domestic and foreign investment capital, and environmental considerations.

Cuba also faces uncertainty with respect to the ability and willingness of Venezuela to continue to subsidize the island. If the subsidies end or are reduced, Cuba will find it difficult to generate the fiscal resources necessary to continue to operate the power system at present levels of generation and tariffs (the cost of electricity).

The MARKAL/TIMES modeling platform represents all energy-producing and -consuming sectors in an integrated and highly transparent framework at a user-specified level of end-use, technology, and pollutant detail. The MARKAL/TIMES energy systems analysis toolbox is well suited to examine interlocking uncertainties through a systematic approach, and includes a model to analyze future capacity investment decisions designed to increase the efficiency of the power system. This modeling paradigm has been in use for more than thirty years at more than two hundred institutions worldwide. The platform has become one of the leading energy-systems modeling frameworks currently in use for several major international and global applications, and for national strategic planning in dozens of developed and developing countries.[13]

In the interest of simplicity and short model construction time, the Cuba MARKAL/TIMES model development and analysis effort focused on the supply and power sectors only, to show electricity demand growth in a simple summary fashion. This section summarizes the model's results.

The goal of the analysis was to identify cost-effective power-sector investment options under various scenarios of electricity demand growth, oil and gas production, and other key energy-system variables over the 2007–25 period. The key uncertainties were divided into two scenario sets, or "storylines." The first is a business-as-usual case that assumes continued moderate electricity load, limited foreign investment in the oil and gas sector, and hence, limited production growth. The second, high-investment case assumes rapid economic and electricity demand growth, high foreign investment, rapid increase in domestic fuel production, and transition to market pricing of electricity. Within each scenario set, sensitivity analyses were conducted on key variables, including higher gas prices, lower oil prices, restrictions on the feasible rate of investment in new power plant and liquefied natural gas (LNG) import infrastructure, and high bagasse availability owing to a revitalized sugar and ethanol industry (bagasse is the fibrous residue left after sugarcane is crushed to extract juice). The scenarios are summarized in table 3-11.

The key results were found to be robust across these multidimensional hypotheticals. Natural gas was found to be the most cost-effective fuel, even in cases where natural gas prices were increased 40 percent above AEO2009 (annual energy outlook for 2009) projected levels and oil prices were set at 40 percent below AEO2009 levels. The model found it to be most cost-effective to replace the existing heavy fuel oil power plants with new natural gas combined-cycle plants as quickly as possible, as a result of low efficiencies, low availabilities, and high maintenance costs of the existing plants. Sensitivity analysis found this conclusion to be robust even when assumed maintenance costs were reduced by half.[14]

Given these results, the key uncertainty is access to natural gas through domestic production and imported liquid natural gas. Generic assumptions

Table 3-11. Development Scenarios and Sensitivity Cases

Development scenario	Characterized by	Sensitivity cases
Business as usual	Moderate electricity demand growth in line with recent trends	Higher gas prices
	Slow growth in oil and gas production in line with recent trends	Lower oil prices
		Limited rate of new investment
		No liquid natural gas imports available
		Higher bagasse volume available
High-investment, high-growth	Rapid economic growth	Higher gas prices
	High foreign investment	Lower oil prices
	Accelerated oil and gas production growth	Limited rate of new investment
	Transition to market electricity prices	No liquid natural gas imports available
		Higher bagasse volume available

Source: Author's compilation based on Manuel Cereijo, "Cuba's Power Sector: 1998 to 2008," and Oficina Nacional de Estadísticas de Cuba, *Anuario Estadístico de Cuba 2008.*

from the U.S. Energy Information Administration were used to characterize LNG import costs. Under these assumptions, importing LNG was found to be cost-effective in every scenario, suggesting that further examination of site-specific LNG infrastructure costs is an important area for future study. Restricting LNG imports substantially increased system costs in every scenario. When LNG import was denied as a model option, wind and sugarcane bagasse resources were used, although they played a marginal role in the nonrestricted scenarios—those in which the ability of Cuba to finance new natural gas plants and a regasification plant is not constrained.

A second key uncertainty is how rapidly existing plants can be replaced. In the high-growth, high-investment case, the unconstrained system built nearly 3 gigawatts of gas combined-cycle generation in the 2012–14 period, at a cost of $2.5 billion. This rate of construction and investment may be unrealistic because of both economic and physical construction constraints. Various restrictions on the speed of this replacement were imposed, and were found to be a primary determinant of system costs and electricity prices.

Current electricity tariffs are subsidized, at an estimated average cost of $0.13 per kWh, or somewhat over half of current costs, assuming an exchange rate of approximately 1.75 CUP = $1. A variety of scenarios endogenizing demand response were explored.[15] The effect was primarily on medium-term demand. Once the system has completed the replacement of existing plants, subsidies are no longer needed to maintain electricity prices at or below current levels until the final periods of the modeling horizon.

In the business-as-usual scenario, limited growth in domestic gas production made the system very sensitive to changes in external conditions. A 40 percent increase in gas prices increased system cost by 15 percent, and an inability to import liquid natural gas increased system cost by 50 percent. With LNG restricted, the system retained the more efficient existing heavy fuel oil plants and built a variety of new plants burning heavy fuel oil to produce steam, wind farms, and bagasse-burning plants. When, in the unconstrained scenario, the rate of new investment was held below the 2.5 gigawatts projected to be built between 2012 and 2014, existing heavy fuel oil plants continued to operate for some time in the presence of the slower turnover.

Fuel expenditures sharply increased in 2010, as Venezuelan oil subsidies were assumed to expire. Switching to natural-gas-fired plants enabled a precipitous drop in fuel costs, even in the high-gas-price case. Limiting access to imported liquid natural gas greatly increased fuel costs, as reliance on petroleum-fired generation was extended. However, a revitalized sugar-bagasse-to-energy industry mitigated this effect by eliminating the need for imported petroleum by 2020.

In the high-investment scenario, system costs were also sensitive to increases in gas prices and restrictions on the rate of investment in new power plants, but the effect was less pronounced than in the business-as-usual case. When electricity subsidies were removed, demand growth moderated, but only slightly. The effect was most pronounced when replacement of the existing plants was slowed by investment constraints. Once the transition was completed, relatively low market prices led to only minor adjustments in demand. The only factors that significantly increased prices beyond this level were restricting new investment or denying LNG imports. Unlike in the business-as-usual scenario, in the high-investment scenario, rapidly expanding domestic gas production enabled system reliance on gas even in the absence of LNG imports, with a minor role for wind and bagasse. However, higher costs led to a downward adjustment of demand in response to market prices, suggesting that end-use energy efficiency potential and cost are important variables for further analysis.

As in business-as-usual, the switch to natural gas in the high-investment scenario enabled a substantial drop in fuel costs. However, higher demand growth left the system vulnerable to increases in natural gas prices, which substantially increased generation costs. Greater domestic production of natural gas than in business-as-usual allowed the system to obtain the majority of its fuel from domestic sources, decreasing energy security concerns.

In both scenarios, replacement of the existing plants led to a sharp drop in CO_2 emissions. By 2025, however, steady growth in generation brought emissions back nearly to 2007 levels (in the business-as-usual case) or above (in the high-investment case). Only in the renewables-heavy, LNG-restricted business-as-usual scenario were CO_2 emissions flat across the time horizon. Scenarios with CO_2 emissions prices or caps were not examined in this analysis, but such an analysis could identify opportunities for emissions reductions.

The analysis suggests the following key areas for further study:

—The cost and potential for future access to natural gas through domestic production and LNG imports

—The feasible rate of power plant replacement

—The potential for price-responsive demand adjustment through end-use energy efficiency

—The feasibility of coal generation

The Potential for Renewables in Cuba

The potential for renewable energy sources is somewhat limited, and more research is needed. Some of the conclusions of the International Resources Group report on the potential for renewables are outlined.

Hydropower

The total hydropower resource in Cuba has been estimated at 650 megawatts,[16] but much of the currently unutilized potential is in protected or naturally sensitive areas that may not be candidates for development. The remaining resource appears suited for small facilities in areas that are mountainous or have seasonal characteristics. Thus, we assumed that these resources could continue to be exploited primarily for off-grid electricity supply to rural schools, medical centers, and small villages rather than for the grid-connected demand considered in this study. Therefore, new hydropower installations were not included in the set of new power plant options.

Solar Photovoltaic

Cuba obviously has excellent solar resources—the use of solar photovoltaic generation is limited by capital cost rather than resource base. We investigated concentrated solar—the use of lenses or mirrors to focus a large area of sunlight onto a small area, whence this concentrated light is directed onto a photovoltaic surface. We did not regard concentrated solar as a realistic grid-connected option, because of limitations in radiation and atmospheric clarity: Cuba's high ambient humidity results in low efficiencies for concentrated solar energy generation. Present photovoltaic applications are largely for off-grid uses. We developed solar PV capacity factors for Cuba from the RETscreen Clean Energy Project Analysis Software, a decision support tool developed with the contribution of numerous experts from government, industry, and academia.[17]

Wind

A preliminary estimate of Cuba's wind potential is 400 megawatts.[18] Wind capacity factors and transmission costs are always highly site-dependent, so only general estimates could be made until a detailed site inventory for Cuba was undertaken. A detailed high-resolution wind energy resource map for Cuba was created at the United States Department of Energy's National Renewable Energy Laboratory as part of the Solar and Wind Energy Resource Assessment project for the United Nations Environment Program. The wind mapping—which used a combination of analytical, numerical, and empirical methods employing GIS mapping tools and data sets—covered approximately 110,000 square kilometers of land area and, when offshore areas were included, more than 150,000 square kilometers. The resulting map highlights the major wind resource areas and provides a wind resource estimate consistent with available measurement data. The report estimated the total electricity-generating wind potential for Cuba at 2,550 megawatts for class 4 and 5 wind areas.[19]

The national wind resource map indicates mostly moderate resources; the largest area with good winds is an area offshore from Guantánamo. Mountain ridges are likely to have small localized good to excellent wind resources.

Biomass

Estimates of the sugarcane bagasse resource in Cuba were derived from work by Walfrido Alonso-Pippo in which he examined the history, methods, costs, and future prospects of Cuba's attempts to develop the energy potential of sugarcane.[20] The paper shows that sugarcane production in Cuba was historically over 70 million tons per year until the early 1990s, when production dropped dramatically, to about 35 million tons per year; it has continued a slow decline since then to about 25 million tons per year in 2009.

Table 3-12. Projections of Annual Bagasse Volume under Two Assumptions of Annual Sugarcane Production
Million tons of bagasse annually[a]

Scenario	2007	2010	2013	2016	2019	2022	2025
Low bagasse volume	25	25	25	25	25	25	25
High bagasse volume	25	30.8	36.7	42.5	48.3	54.2	60

Source: Author's compilation based on Manuel Cereijo, "Cuba's Power Sector: 1998 to 2008," and Oficina Nacional de Estadísticas de Cuba, *Anuario Estadístico de Cuba 2008.*
a. The energy content of one ton of bagasse is 2.3 petajoules.

From the 1990s to the present, the Cuban sugarcane industry's average sugarcane yield has declined from 57.5 to 22.4 tons per hectare in 2005, a 39 percent decrease. At the same time, the total amount of agricultural land in Cuba used for sugarcane cultivation has declined from 21 percent to barely 5 percent. Alonso-Pippo cites many reasons for the decline of Cuba's sugarcane industry and for the failure of the industry to recover in spite of government efforts to improve it. Despite these factors there is significant room for expanding cane production if sufficient investment and improved incentives were employed. Jorge Piñón, a former oil industry executive, puts forth two scenarios for the resource potential of sugarcane yields, one based on the current sugar industry and one based on a revitalized sugar-ethanol-bagasse industry (see table 3-12).

The bagasse, the biomass that remains at the mills after the sugarcane has been ground and crushed to extract the juice, represents only about 15 percent of the weight of the dry sugarcane. The average heat content of sugarcane bagasse is approximately 16.5 gigajoules per dry ton.[21] The potential energy available from Cuba's bagasse resource is given in table 3-13 for each of the two scenarios.

Estimated electricity production from bagasse at sugar-ethanol mills was based on the operation of a 7,000-ton-per-day sugar mill producing either sugar or ethanol. The crushing season is about three thousand hours per year.

Table 3-13. Bagasse Energy Potential in Two Scenarios

Scenario	2007	2010	2013	2016	2019	2022	2025
Low bagasse	58.5	58.5	58.5	58.5	58.5	58.5	58.5
High bagasse	58.5	72.2	85.8	99.5	113.1	126.8	140.4

Source: Author's compilation.

Table 3-14. Technology Characterization for Bagasse Plant, Capital, and Operating Costs

Capital investment	U.S.$ million
Steam saving from sugar factory	3.3
Distillery	6.4
Cogeneration capacity (33 megawatts)	50.4
Total investment	60.1
Operations and management costs (U.S.$ per kilowatt-hour)	
Fixed	0.031
Variable	0.015

Sources: Based on Amy Myers Jaffe and Ronald Soligo, *The Potential for the U.S. Energy Sector in Cuba* (Rice University, 2001) (www.cubafoundation.org/CPF-EnergyStudy.htm), and J. R. Piñón Cervera, "Cuba's Energy Challenge: A Second Look," Association for the Study of the Cuban Economy, Miami, August 2–4, 2005.

The calculation of the amount of surplus electricity that can be cogenerated is premised on efficiency improvements to reduce steam use in the sugar factory, the addition of a distillery to increase revenue, and the addition of a 33 megawatt cogeneration plant that uses a condensing-extraction steam turbine. The plant generates 92 gigawatts per year of surplus power, which is about a 32 percent capacity factor—the ratio of the actual output of a power plant over a period of time and its output if it had operated at full nameplate capacity the entire time. The capital investment cost for the steam reduction, the distillery, the 33 megawatt cogeneration plant, and the operating and maintenance costs are shown in table 3-14.

Increasing the Efficiency and Sustainability Potential of the Power Sector

No major changes would be required to achieve the "business as usual" scenario. Achieving the high-growth and high-investment scenario would require implementing different policy measures, including the rationalization of the tariff regime. Equally important would be establishing a more favorable environment for foreign investment; this is a particular challenge, as the general environment for foreign direct investment is being affected negatively by the global financial crisis.

What measures should be implemented to achieve the high-investment scenario? How can the goal be met of attracting investment of $2.5 billion for generation using combined-cycle gas turbines, plus the resources necessary to expand domestic gas production or to establish a regasification plant to handle imported liquid natural gas? As it strives to implement policies necessary to achieve the high-investment scenario, Cuba could benefit from the experience of other Latin American, Asian, Eastern European, and Central Asian nations.[22] But first, some caveats:

—These recommendations are not offered prescriptively but rather to highlight some of the main issues that will have to be faced if a Cuban government makes the decision to reform the country's power sector. The verbs "should" and "would" are used merely as shorthand, to indicate that authorities in Cuba should consider adopting the proposed measures.

—Power-sector reforms can only succeed if they are coordinated with the policies guiding reforms in other sectors. That is, tariff adjustment, protection of vulnerable groups, and private participation policies must be formulated in the context of national policies on those subjects.

—An adequate power supply is a necessary condition for rapid economic growth, so decisions on reforming the power sector are extremely important. These decisions must be made autonomously by local authorities vested with such decisionmaking responsibility, but the experiences of other countries can inform the direction taken by Cuba, and international experts can provide valuable advice.

—Models from other countries should not be applied wholesale in Cuba but rather must be adapted to local conditions.

—Most important, our recommendations rest on the premise that the Cuban government has made the decision to improve the environment for foreign investment in the power sector. Support of the U.S. government, though not absolutely indispensable, could help make Cuba's transition to cleaner and cheaper energy faster and easier. In this context, the recommendations of the recently completed report by a staff member of the Senate Committee on Foreign Relations are particularly relevant. He concluded that the United States should increase energy cooperation with Cuba.

Models of Power-Sector Reform

Most of the literature on power-sector reform advocates that countries with a sufficiently large market should move to a fully competitive power system.[23] (A rule of thumb is that competition is possible in markets selling above 1,000 megawatts of installed capacity.) A way to understand the path to reform is to introduce the concept of power-sector models, which can be summarized as follows:[24]

Monopoly (Model 1)

In this model there is no competition at all levels of the supply chain (generation, transmission, distribution, and commercialization), that is, a single

Figure 3-5. Structure of Cuba's Power Sector, 2010

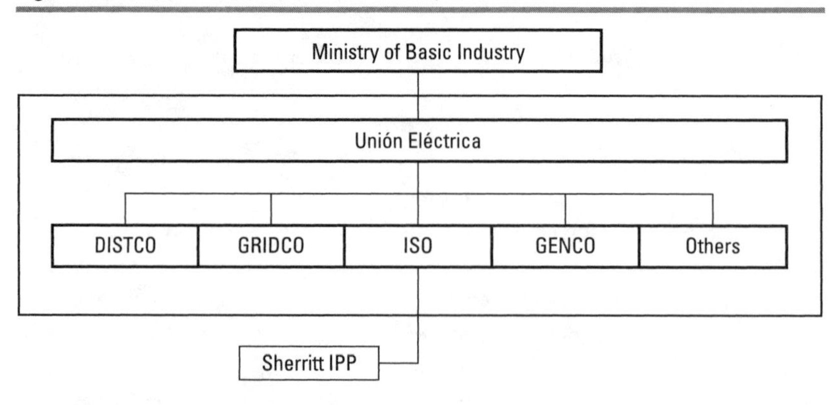

Source: Author's compilation.
DISTCO = distribution company.
GRIDCO = high-voltage transmission grid.
ISO = independent system operator.
GENCO = generation company.
Others = support activities (construction personnel and other assets).

company produces and delivers electricity to the final users. These monopolies can be private or state-owned. In Cuba today, Unión Eléctrica essentially dominates all aspects of the power sector. This model best represented the power sector in Cuba until the power purchase agreements with Sherritt were introduced.

Single Purchasing Agency (Model 2)

In this model a single buyer or purchasing agency buys electricity from a number of different generators, normally referred to as independent power producers (IPPs), with whom it has purchase contracts. These power purchase agreements (PPAs) are reached either through bilateral negotiations or through a competitive process. In quite a few countries, El Salvador and Guatemala, for example, the reform of the power sector began with a state-owned enterprise signing PPAs with IPPs. While these purchase agreements fulfilled the objective of supplying emergency needs, as the frequency of blackouts was reduced, some have criticized them as being costly, since many of these contracts were the result of bilateral negotiations rather than the result of competitive bidding processes. In systems that introduced competition in generation, the PPAs signed prior to the reforms represent "stranded costs" that are being paid by consumers. Cuba today follows a very limited version of model 2—it could be termed model 2 "lite" (see figure 3-5).

Wholesale Competition (Model 3)

This model represents a departure from the traditional way to manage power systems. Basically, an original monopoly is separated both horizontally and vertically. This separation, called unbundling, results in a sector structure with several generators (generating companies, or GENCOS), several distributing companies (DISTCOS), one or more transmission companies (TRANSCOS), and a dispatch center. DISTCOS purchase electricity directly from the generators they choose, transmit this electricity under open-access arrangements over the transmission system to their service area, and deliver it over their local grids to their customers. The regulator allows competition in generation but regulates the value added in distribution (VAD), that is, the markup between the wholesale cost of the electricity and the final price paid by the consumers. Regulation of the VAD is normally through a price cap.

Partial Retail Competition (Model 4)

This model is like model 3, but large customers (defined in the regulations) can negotiate prices directly with generators. This adds competition to the system, and also allows large clients with special needs, such as those with peak demand different from that of the average consumer, to negotiate terms that are favorable to them and also to the generator. This model could represent the goal of reform of Cuba's power sector three to five years after the beginning of a reform effort. I have advocated elsewhere that Cuba should eventually move to model 4, which is what predominates in Latin America; reaching that stage would I believe take three to five years after the decision to reform the sector is taken.[25]

Full Retail Competition (Model 5)

Under model 5, all customers can choose their electricity supplier, under open access for suppliers to the transmission and distribution systems. This model has been implemented in California, a market design some argue was unduly influenced by ENRON.[26]

Preconditions to Modernizing the Cuban Power Sector

Here I shall identify the different measures that the government of Cuba should consider to strengthen the present power-sector model (model 2 in the literature on power-sector reform) in order to encourage foreign direct investment in generation facilities and technology. The proposed future

Figure 3-6. Structure of Cuba's Energy Sector to Achieve the High-Growth, High-Investment Scenario

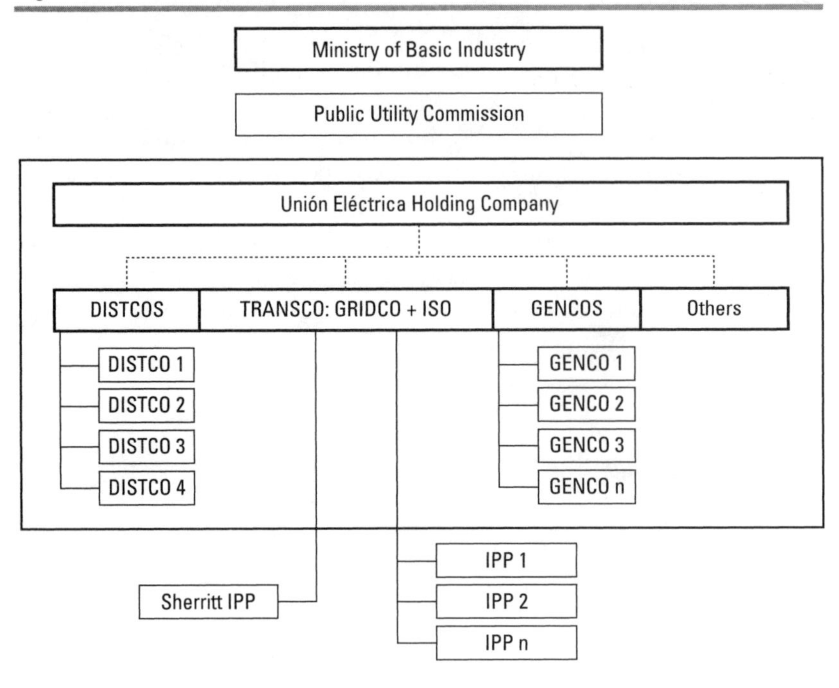

Source: Author, based on proposed market reform model.
DISTCO = distribution company.
TRANSCO = transmission company.
GRIDCO = high-voltage transmission grid.
ISO = system operator.
GENCO = generation company.
Others = support activities (construction personnel and other assets).

structure necessary to achieve the high-growth and high-foreign-investment scenario is shown in figure 3-6.

Some of the steps that should be implemented to achieve the high-growth, high-investment scenario include the following:

—Pass electricity-sector legislation and establish a public utility commission

—Model the energy sector to determine optimal expansion path

—Modify tariffs so as to reach full cost recovery

—Restructure Unión Eléctrica

—Develop independent power producer arrangements

—Develop operating contracts or concessions for existing assets

Establish a Public Utility Commission

The electricity-sector law ("the law") should require the Ministry of Basic Industry (Minbas) to concentrate its efforts in the power sector on the formulation of strategy and policies, while a public utilities commission (PUC) should be established to regulate the sector. Latin American countries that have reformed their power sectors have learned the basic lesson that it is easier to develop a legal and regulatory framework than to develop adequate institutions, such as relevant ministries and regulatory commissions, whose task is policy formulation. The regulatory agencies—the public utility commission—must be independent of the former and must be accountable to a board of commissioners for its actions.

An effective autonomous regulatory public utilities commission, as set forth by Warrick Smith, has the following features:[27]

—Established by law

—Has arm's-length relationship with operators, consumers, and other private interests

—Has arm's-length relationship with political authorities

—Is financially independent (financed by a fee charged to the regulated companies) and can pay commissioners competitive salaries

—Commissioners chosen with participation of executive and legislative branches of government

—Professional criteria used for appointment of commissioner(s)

—Commissioners have fixed, staggered terms and can be removed only for well-defined cause

Independence must be balanced by accountability, which is ensured by the following:

—Strong provisions prohibiting conflicts of interest

—Established rules and procedures for appealing decisions of the regulator

—Public availability of budget and scrutiny by (usually) Parliament

—External audits

—Permitting removal of commissioner(s) for just cause

—Open hearings with participation by the regulated industries and the consumers

The staff of the public utility commission will require training on the job as well as abroad. Good possibilities for training abroad include the Public Utilities Research Center of the University of Florida, the Kennedy School of Government at Harvard (specifically, the course "Infrastructure in a Market

Economy"), and the Institute for Public-Private Partnerships, headquartered in Arlington, Virginia. Foreign consultants could provide on-the-job training and also support the PUC. Another powerful instrument for enhancing the skills of the staff of the Cuban regulatory commission would be partnerships, under the National Association of Regulatory Utility Commissions, with a U.S. state regulatory body. USAID is supporting a partnership between the regulator of Nicaragua and the regulator of Texas, and the results have been positive.[28]

Model the Energy Sector

Modeling the Cuban energy sector means assessing what would be necessary to change the power sector from model X to model Y. Ideally it would done in concert with Unión Eléctrica technicians, who would have the most accurate data and the greatest knowledge of the power system. A thorough modeling of the whole energy sector (power, transport, industry, and households) and the generation of more, and more accurate, data could help determine with greater precision the optimum path for restructuring the power system. To gain a better overview of the total energy sector, additional studies should be undertaken to provide the following:

—A more thorough analysis of the prospects for renewable energy

—The potential for and cost of future access to natural gas through domestic production and imported liquid natural gas

—Prefeasibility study for building a degasification plant and/or domestic gas transmission

—Modeling of the transmission network to determine the optimal location of power plants, including combined-cycle gas turbines

—Determination of the feasible rate of power plant replacement

—Potential for price-responsive demand adjustment through end-use energy efficiency

Modify Tariffs

A new tariff schedule should be developed that reflects new cost estimates and increases in efficiency. The preliminary analysis shows that if the sector is transformed from liquid fuels to gas and renewables, there may not be a need for major tariff increases. Any tariff changes should be established by the public utility commission and a system designed to adjust prices to reflect changes in costs be put in place.

Table 3-15. Two Scenarios for Creating Power Distribution Entities

Number of distribution areas	Province or region	Estimated number of customers (millions)	Market participation (percent)
Three distribution areas	Havana province	1.1	33.3
	Central provinces	1.2	36.4
	Eastern provinces	1.0	30.3
	Total	3.3	100
Four distribution areas	Havana province	1.1	33.3
	Central provinces, region 1	0.7	21.0
	Central provinces, region 2	0.8	22.8
	Eastern provinces	0.8	22.9
	Total	3.3	100

Source: Author's compilation based on Manuel Cereijo, "Cuba's Power Sector: 1998 to 2008," and Oficina Nacional de Estadísticas de Cuba, *Anuario Estadístico de Cuba 2008.*

Restructure Unión Eléctrica

The Cuban government should consider unbundling Unión Eléctrica by establishing separate entities according to function, under a holding company (see figure 3-6). For example, the distribution assets could be separated into three or four separate distribution companies (see table 3-15 for the possible division of those assets). Planners would need to develop a more thorough analysis of the different load centers and of the transmission network to ensure a rational, appropriate separation according to actual power markets.

Sally Hunt recommends that the system's operation and the transmission should be combined and corporatized into a combined operations and transmission company, or TRANSCO.[29] Hunt's reasoning is that combining systems operation and transmission provides a better business model and would improve coordination of power supply and demand. Transmission and system operations have been combined effectively in a number of markets, including England and Wales, Spain, and Scandinavia.[30] Some argue against this model because it concentrates too much power in one institution.

The Cuban government should consider breaking up the generation assets of Unión Eléctrica and establishing seven to eight generating companies. Additionally, a company could be established with the personnel and assets to carry out construction.

Develop Independent Power Producer Contracts

Using the results of the planning model, a transparent system should be developed for independent power producer (IPP) life-of-plant contracts

(contracts that last as long as the useful life of the asset). Some key questions need to be answered concerning the design of these contracts:

—Should the technology to be used be specified?

—Should the contracts be outside the control of the system operator?

—Should the profits be calculated as a function of the fixed assets, or the variable costs, or a combination of both?

Develop Operating Contracts and/or Concessions for Existing Assets

Cuba has demonstrated that it is ready to permit significant private participation in the provision of infrastructure services. Besides its IPP contract with Energas-Sherritt, Cuba has privatized most of the telecommunications company and developed concessions for the water utilities in Havana and Varadero, with Aguas de Barcelona and Aguas de Valencia, respectively. Cuba should also consider introducing operating contracts (incentive-based management contracts) and/or concessions for some of the state-owned generation and distribution companies.

Support by the U.S. government for the potential reforms of the power sector is severely constrained by existing Helms-Burton legislation. If legislation is modified or a waiver is granted, the American government could support technical assistance for modeling the sector to determine the potential for improving efficiency and environmental sustainability and for training government officials on economic regulation of utilities. Training possibilities include:

—Course on regulation at the Public Utilities Research Center at the University of Florida. This course takes place twice a year, in June and January.

—Course on private infrastructure at the Kennedy School of Government, Harvard University. This course is offered every summer.

—Courses on the design of IPPs and rate design at the Institute for Public Private Partnerships (IP3), which are offered several times each year.

Greater cooperation between the United States and Cuba on the energy sector, including the power sector, would present an opportunity for an enhanced dialogue that could benefit both countries.

Notes

1. Vladimir Ilyich Lenin, "Report on the Work of the Council of People's Commissars," December 22, 1920, *Collected Works,* vol. 31. Cited in the *Columbia World of Quotations* (Columbia University Press, 1996).

2. The conclusion that Cuba has a higher consumption of electricity is based on a regression of the natural logarithm of electricity consumption as a function of the natural logarithm of GDP per capita (PPP basis). Calculation by author based on data collected by the Oficina Nacional de Estadísticas de Cuba for electricity consumption and presented in *Anuario Estadistico de Cuba 2008,* Oficina Nacional de Estadísticas, 2009, table 10-13.

3. In 2003 Cuba produced 93 percent of its power using liquid fuels, the fifth-highest percentage in the world. The decline in that percentage is the result of new gas-fired facilities established under a power purchase agreement with Sherritt, a Canadian company.

4. Manuel Cereijo, "Cuba's Power Sector: 1998 to 2008," paper presented at the 18th annual meetings of the Association for the Study of the Cuban Economy, Miami (August 2008).

5. Simon Romero, "In Cuba, a Politically Incorrect Love of the Frigidaire," *New York Times,* September 2, 2007, p. A1.

6. Cereijo, "Cuba's Power Sector: 1998 to 2008," 28.

7. Financial records for Petróleos de Venezuela are also unavailable.

8. I am grateful to Jorge Sanguinetty, who raised this issue in a conversation.

9. Cereijo, "Cuba's Power Sector."

10. Officially, Cuba buys oil from Venezuela at a 40 percent discount—roughly the proportion of domestic oil that belongs to Cuba under the production-sharing agreement with foreign oil companies producing in Cuba.

11. In Cuba, energy theft is penalized with exorbitant fines and even prison. Energy theft generally is done through meter tampering or by connecting illegally to the distribution network.

12. This Section draws on Evelyn Wright and others, "A Power Sector Analysis for Cuba Using the Markal Model," *Cuba in Transition* 19: 493–96 (http://lanic.utexas.edu/project/asce/pdfs/volume19/pdfs/wrightbeltetal.pdf). This analysis was hindered by the lack of reliable data. See next note.

13. MARKAL/TIMES is an energy systems modeling platform used to construct models and analyze energy, economic, and environmental issues over several decades. This set of software tools provides a framework for exploring, evaluating, and quantifying alternative energy futures and the roles that various policy options may have on technology and resource choices. For information on MARKAL/TIMES, please see www.irgltd.com/Our_Work/Services/MARKAL_TIMES%20Demystified-v02.pdf. To understand the breadth of the MARKAL/TIMES model's applications, please see, among many other examples:

—www.sofreco.com/projets/c886/Reports.htm

—http://pesd.stanford.edu/news/chinagasreport

—www.ukerc.ac.uk/ResearchProgrammes/EnergySystemsandModelling/ESM.aspx

—http://fsi.stanford.edu/publications/20219/

—www.nrdc.org/media/2008/080513.asp.

14. Future analysis should also look more closely at the economic and financial viability of coal. The government of Cuba recently announced a plan to establish a gas regasification facility in the port of Cienfuegos. If this facility were built, gas would become the most viable source of power.

15. Endogenizing demand means allowing demand to decline in response to price increases.

16. International Solar Energy Society, Sustainable Energy Policy Concepts (SEPCO) website, Country Case Study Cuba, "Development of Cuba's Energy Supply in the Last Decade," 2005 (www.ises.org/sepconew/Pages/CountryCaseStudyCu/2.html); D. Perez, I. Lopez, and I. Berdellans, "Evaluation of Energy Policy in Cuba Using the Indicators for Sustainable Energy Development (ISED)," *Natural Resources Forum* 29 (2005): 298–307. Since 1999 the International Atomic Energy Agency (IAEA) has been leading a multinational, multi-agency effort to develop a set of energy indicators useful for measuring progress on sustainable development at the national level. This effort has included the identification of major relevant energy indicators and the development of a framework for implementation and the testing of the applicability of this tool in a number of countries. To achieve these goals, the IAEA has worked closely with other international organizations that are leaders in energy and environmental statistics and analysis, including the United Nations Department of Economic and Social Affairs, the International Energy Agency, Eurostat, and the European Environment Agency. Also, the IAEA completed a three-year coordinated research project for the implementation and testing of the original set of indicators in seven countries: Brazil, Cuba, Lithuania, Mexico, the Russian Federation, the Slovak Republic, and Thailand.

17. RETScreen Clean Energy Project Analysis Software (www.retscreen.net/ang/home.php) provided by Natural Resources Canada.

18. Perez, Lopez, and Berdellans, "Evaluation of Energy Policy in Cuba Using ISED."

19. United Nations Environmental Program, Solar and Wind Energy Resource Assessment, "Cuba Wind Energy Resource Mapping Activity," technical report, August, 21, 2006, p. 9 (www.fishermensenergy.com/dms/showfile.php?id=206).

20. Walfrido Alonso-Pippo and others, "Sugarcane Energy Use: The Cuban Case," *Energy Policy* 36 (2008): 2163–81.

21. Environmental Protection Agency, "Bagasse Combustion in Sugar Mills," background document no. AP-42, section 1.8 (www.epa.gov/ttn/chief/ap42/ch01/final/c01s08.pdf).

22. See Juan A. B. Belt, "Telecom and Power Sector Reforms in Latin America—Lessons Learned," *Cuba in Transition* 10 (2000): 374–81; Juan A. B. Belt, "Power Sector Reforms in Market and Transition Economies—Lessons for Cuba," *Cuba in Transition* 16 (2006): 75–88; Juan A. B. Belt and Luis Velasquez, "Cuba: Reforming the Power, Telecommunications and Water Sectors during a Transition," *Cuba in Transition* 17 (2007): 59–75.

23. Robert Bacon, "Appropriate Restructuring Strategies for the Power Generation Sector: The Case of Small Systems" (Washington: World Bank, 1995).

24. This typology is a simplification, and all models have nuances and "submodels." For example, in some countries, state-owned enterprises have been corporatized and subjected to additional discipline, including hard budget constraints.

25. Belt, "Power Sector Reforms in Market and Transition Economies—Lessons for Cuba," 75–84.

26. Sally Hunt, *Making Electricity Work in Competition* (New York: John Wiley, 2002).

27. Warrick Smith, "Utility Regulators: The Independence Debate," *Public Policy for the Private Sector,* note 127 (Washington: World Bank Group, 1997), 1–4.

28. Personal communication with Ing. José David Castillo Sánchez, head of the regulatory agency of Nicaragua, Erin Skootsky of National Association of Regulatory Utility Commissioners (NARUC), and Timothy O'Hare of the USAID mission in Nicaragua.

29. Hunt, *Making Electricity Work in Competition,* 302–03.

30. Ibid., 212.

four
Energy Balances and the Potential for Biofuels in Cuba

RONALD SOLIGO AND AMY MYERS JAFFE

In 2002 we published a paper on Cuban energy that began with the observation "Cuba is considered a promising growth energy market in the Americas."[1] Eight years later Cuba shows even greater promise in the energy sector, but progress in realizing energy opportunities has been slow. Gaining a better understanding of Cuba's energy potential is important for policymakers in the United States, Cuba, and the Caribbean region. From the American point of view, the possibility of having an additional supplier of energy to the U.S. market located just a few miles offshore could contribute significantly to the United States' energy security. The magnitude of Cuba's energy resources is uncertain, but one estimate, by the U.S. Geological Survey, is that Cuba has mean "undiscovered" reserves of 4.6 billion barrels of conventional oil and 9.8 trillion cubic feet of natural gas in the North Cuba Basin.[2] In addition, Cuba has large land areas that once produced sugar but now lie idle. These could be revived to provide a basis for a world-class ethanol industry. We estimate that if Cuba achieves the yield levels attained in Nicaragua and Brazil and the area planted with sugarcane approaches levels seen in the 1970s and 1980s, Cuba could produce up to 2 billion gallons of sugar-based ethanol per year.

The authors thank Matthew Osher for his help in the preparation of this paper. The authors have also benefited from comments by Jonathan Benjamin-Alvarado and Jorge Piñón. We are especially grateful to our colleague Kenneth Medlock for his assistance in generating the energy forecasts.

Despite this potential, Cuba remains dependent on energy imports on a concessional basis. The collapse of its economy when assistance from the Soviet Union was terminated by that country's breakup demonstrated in a dramatic way the benefits that could accrue if Cuba became energy self-sufficient.

Cuba has been thwarted by U.S. economic sanctions and other internal domestic barriers in aggressively pursuing its own energy independence. For the time being, Havana has adopted a policy of replacing former Soviet energy assistance with current Venezuelan aid. This is a risky strategy, since Venezuelan beneficence is dependent on Caracas's own economic health, which is currently shaky.[3]

In this chapter we argue that given its rich potential in both conventional energy and biofuels, Cuba can be both energy-independent and an energy exporter.

State of the Cuban Economy

The Cuban economy suffered a major economic decline after the 1991 collapse of the Soviet Union, with GDP falling 35 percent between 1990 and 1993.[4] The economy began a slow recovery in the mid-1990s, but the extent of the recovery is disputed. International Energy Agency data show that per capita income in purchasing power parity dollars grew on average about 2 percent per annum between 1994 and 2005. CIA data show an increased growth rate for the period 2005 to 2008 ranging between 4.3 percent and 8 percent.[5] Most recently, in 2009 Cuban growth fell to 1.4 percent.[6]

The higher growth rates of Cuban GDP during the period from 2005 to 2008 reflect the increase in Venezuelan aid as well as Cuba's participation in the worldwide commodity boom and some increase in foreign investment.

The extent of Venezuelan aid is not fully transparent. Cuba is importing around 92,000 barrels a day (b/d) from Venezuela under favorable terms. Some of the oil is financed by loans, part is a barter trade involving some 20,000 Cuban medical professionals who work in Venezuela, and some oil is provided as an outright grant. Venezuela has also financed the completion of the Cienfuegos refinery, which was opened at the end of December 2007, with plans to increase the plant's capacity from around 65,000 b/d to 150,000 b/d.[7] Venezuela has also provided $122 million to finance the acquisition of tankers to carry Venezuela crude and products to Cuba.

The commodity boom has also helped the Cuban economy. Cuba is among the top six or seven largest producers of nickel in the world. Production and export of nickel rose from 26.9 million metric tons (mmt) in 1993 to roughly 76 mmt in 2004. Output has subsequently stagnated, but new investment and capacity expansion are on the way. Nickel prices rose from $15,000/ton in 2004–05 to over $50,000/ton in early 2007. According to United Nations data, exports of nickel oxide in 2005 and 2006 each contributed over a billion dollars to Cuba's export earnings.[8] Earnings soared in 2007, with the run-up in nickel prices making nickel a close contender with tourism as Cuba's top export earner. With the onset of the worldwide recession, prices fell and fluctuated between $25,000 and $35,000/ton before plunging below $10,000 in early 2009. The collapse of nickel prices has seriously affected Cuban capacity to import and sustain per capita GDP.

Large foreign investments, in both the energy and non-energy sectors, have been announced in recent years, but it is difficult to estimate actual foreign investment, since there is always a discrepancy between announcements of investment plans and their actual completion. Cuba has not published data on foreign direct investment since 2001.[9]

In addition to Venezuela's investments, some significant projects include a Brazilian loan of up to $450 million to rebuild the port at Mariel to handle larger ships—possibly in anticipation of the end of the U.S. embargo.[10] An end to U.S. economic sanctions would allow Cuba to become an important trans-shipping hub serving overcrowded U.S. ports, possibly sidestepping U.S. environmental restrictions. (The end of U.S. economic sanctions would also open the possibility that Cuba's petroleum-refining industry could export surplus refined products to the U.S. market.) China has announced plans to invest $500 million in Cuba's nickel mining industry.[11] And Russia's Inter RAO UES, an energy company, has recently agreed to form a partnership with Cuba's Unión Eléctrica to upgrade some electricity-generating plants.[12]

Cuban growth has stagnated since 2009 for a number of reasons, most of them related to the worldwide economic recession. The collapse of nickel prices along with a drop in spending by tourists has severely reduced foreign exchange earnings. Foreign investments have been postponed, including those in the energy sector. In addition, several major hurricanes in 2008 diverted resources away from development to relief efforts and reconstruction.

The Energy Sector and Foreign Investment

In the energy area, some limited progress has been made in developing Cuba's offshore oil and gas deposits. Cuba produced about 52,000 barrels per day of oil in 2008, of which Sherritt International, a Canadian company, had an average working interest of around 31,200 b/d.[13] Another Canadian company, Pebercan, was also active in Cuba, producing 19,000 b/d in 2008, but its production-sharing agreement with Cupet, the state-owned Cuban oil company, was terminated in early 2009. Although most of Cuba's oil potential lies farther offshore, current production comes from shallow water just off Cuba's coast, in some cases using slant drilling technology to reach deposits that lie farther off the coast. For example, the 2004 discovery at Santa Cruz has a well bore four kilometers long that extends three kilometers into the Bay of Cardenas.

Cuba also produces 3.45 million cubic meters of associated gas per day. Traditionally the gas has been piped to Havana for commercial and residential use, but with increased production, gas is now also used to produce electricity by Energas, a joint venture consisting of Sherritt, Cupet, and Unión Eléctrica, the state-owned electricity company. Energas has 376 megawatts of capacity, which is 12 percent of Cuban electricity consumption. An expansion of an additional 150 megawatts had been scheduled to begin in 2008, but has been suspended while the work schedule is being reviewed.[14]

In July 2004 Spain's Repsol-YPF identified five "high-quality" fields in the deep water of the Florida Strait, twenty miles northeast of Havana. Cuba has offered fifty-nine new exploration blocks in the area for foreign participation. Several foreign companies, including Venezuela's PDVSA (Petróleos de Venezuela S.A.), have expressed interest in joining exploration efforts in Cuba. A consortium of Repsol-YPF, Norway's Statoil–Norsk Hydro, and India's ONGC Videsh has announced plans to commence drilling a number of times. The latest projection is that drilling will begin in 2011.[15] PetroVietnam and Sinopec (China Petroleum and Chemical Corporation), the Chinese state-owned oil company, have also signed cooperation agreements with Cupet. Brazil's Petrobras, too, has indicated that it will undertake exploration in deepwater areas.[16]

The pursuit of deepwater offshore deposits may not unfold as rapidly as the news releases suggest. Exploration in these areas has been limited in the past because the technology needed to explore and develop deepwater deposits was owned by international oil companies that were severely constrained by U.S. sanctions. Today, other companies, such as Petrobras and Norsk Hydro, have the technology, but they are still reluctant to challenge U.S. sanctions. In addition, under U.S. law, ships that visit Cuban ports are barred from U.S. ports for a period of six months. If this policy is applied to drill ships, as is most likely, it will impose higher costs on oil companies. Drill ships can earn several hundred thousand dollars a day, so each day used to move them from one location to another is costly. By denying immediate access to the U.S. Gulf, the policy forces companies to move their drilling vessels to more distant locations driving up the cost of using them in Cuba. The absence of markets for services, equipment, and supplies in Cuba itself adds to the difficulty and cost of mounting a serious exploration and production effort, because oil firms must plan and bring all equipment and other necessary materials to Cuba rather than rely on local suppliers. Jonathan Benjamin-Alvarado estimates that the absence of these suppliers adds up to 30 percent to a project's cost.[17]

Energy Balances

Cuba's energy profile has changed sharply over the years, reflecting the various periods in its economic history (see table 4-1).[18] Oil remains at the heart of energy supply, accounting for 80 percent of total primary energy supply (TPES) in 2006. Biomass, primarily from sugarcane, has also been important, but its share in TPES has decreased over time with the decline in the sugar industry. Energy supply was higher prior to 1991, because of the generous supplies from the USSR on concessional terms. After the collapse of the Soviet Union, Cuba was forced to buy its supplies in the international market, and TPES, which had fluctuated between 15,000 and 16,800 kilotons of oil equivalent (ktoe) in the preceding decade, dropped to 10,437 ktoe in 1995. Crude supply, which had been fairly consistent at around 6,800 ktoe until 1990, fell dramatically to 2,625 ktoe in 1995. The share of oil in total energy supply, which had fluctuated between 45 percent and 50 percent prior to 1990, also fell to a low of 19 percent in 1992 as Cuba shifted toward importing relatively more refined petroleum products and less crude.

The shift from importing crude to importing refined products reflected the decline in Cuba's refining capacity, the result of poor maintenance exacerbated by the loss of Soviet aid, which reduced Cuba's capacity to finance imports of parts and equipment. Another possible factor was that since Cuba no longer had access to USSR crude oil on a preferential basis, it was forced to buy at market prices. Importing refined oil products may have represented a better value than importing crude and refining it at high domestic cost.

As an aside, it is important to note that TPES measures only the energy consumed in Cuba. In fact, Cuba received much more oil from the USSR than it consumed, re-exporting a significant amount to finance imports of other goods and services. In 1985 these exports amounted to 3,500 ktoe.

The end of Soviet involvement in Cuba also brought sectoral changes in the use of energy. Although all economic sectors were affected, the residen-

Table 4-1. Cuba's Energy Profile
Kilotons of oil equivalent (ktoe)[a]

Sector and source	1975	1985	1990	1995	2000	2006
End sector						
Residential/commercial	880.3	1403.3	1470.7	986.7	1181.1	1244.6
Transportation	2,263.6	2,997.7	1,980.1	865.9	946.5	956.0
Industrial/agricultural/other	6,678.9	6,888.7	8,629.2	4,564.1	5,226.3	3,711.0
Total final consumption (TFC)	9,822.7	11,289.7	12,079.9	6,416.6	7,353.9	5,911.6
Total primary energy supply (TPES)[b]	13,184.3	14,601.7	16,834.7	10,436.8	11,500.8	10,639.0
TFC/TPES	0.745	0.773	0.718	0.615	0.639	0.556
Primary energy source						
Coal	63.4	120.8	137.4	53.0	22.0	23.0
Crude, natural gas liquids, and feedstocks	6,746.3	6,856.4	6,860.3	2,624.6	4,269.1	5,593.2
Petroleum products	2,764.4	3,508.2	3,929.5	4,734.9	3,634.9	2,860.7
Natural gas	14.2	5.7	27.5	14.1	468.8	886.1
Hydro	5.3	4.6	7.8	6.4	7.7	8.1
Combustible renewables and waste	3,590.6	4,106.0	5,872.1	3,003.8	3,098.2	1,267.9
TPES	13,184.3	14,601.7	16,834.7	10,436.8	11,500.8	10,639.0
Domestic oil production	254.6	861.1	659.6	1,446.0	2,649.2	2,850.7
Imports of crude	6,514.3	7,988.8	6,200.7	1,178.6	1,620.0	2,742.6
Energy source as share of total						
Coal	0.5	0.8	0.8	0.5	0.2	0.2
Crude, natural gas liquids, and feedstocks	51.2	47	40.8	25.1	37.1	52.6
Petroleum products	21	24	23.3	45.4	31.6	26.9
Natural gas	0.1	0	0.2	0.1	4.1	8.3
Hydro	0	0	0	0.1	0.1	0.1
Combustible renewables and waste	27.2	28.1	34.9	28.8	26.9	11.9

Source: International Energy Association, IEA Energy Balances, online subscription database (www.iea.org/stats/index.asp).

a. 1 toe (ton of oil equivalent) = 6.449 barrels for Cuba (per IEA).

b. Total primary energy supply is equal to domestic production plus net imports.

tial and commercial sectors were affected proportionately less than industry and agriculture, because the government gave priority to maintaining electrical services to households and commercial establishments. The Cuban government also undertook policies to produce a dramatic reduction in energy use in the transportation sector. Public transport was severely reduced. The number of buses operating in Havana dropped from 2,200 before the crisis to only 500 by 1993. Bus routes outside Havana were cut by two-thirds, and bicycles were imported from China as a substitute mode of transportation.[19] Energy consumption in the transport sector has never recovered from these severe cuts. Total final demand in the transportation sector in 2005 was less than a third of its 1988 level. Similar drastic cuts were undertaken in the industrial-agricultural sector; the collapse of the sugar industry was one source of cuts as energy requirements for growing and processing sugarcane declined sharply. Total primary energy supply fell from 8,629 ktoe in 1990 to only 3,711 in 2006. Although the Cuban economy had begun to recover by the second half of the 1990s, TPES, which had fallen to a low of 10,210 ktoe in 1993, did not significantly increase, averaging only 10,822 ktoe during the period from 1995 to 2006. Crude oil supplies increased from a low of 2,254 ktoe in 1992 to 5,593 ktoe in 2006 as domestic production more than doubled and imports rose. However, this increase in crude supplies was offset by a fall in imports of petroleum products. The balance between the supply of crude and petroleum products shifted back toward its historical ratio, where crude accounted for roughly half of TPES.

In recent years the supply of natural gas, all of which is produced domestically in association with crude oil output, has increased significantly, along with the increased production of oil. Total annual supply of gas averaged 19.1 ktoe between 1990 and 1996 but rose rapidly to 896.1 ktoe in 2006.

A troubling aspect of the Cuban energy profile is the ratio of total final consumption, or TFC, to total primary energy supply, or TPES, which has shown a steady decline since 1990. This ratio is a measure of the energy lost during conversion from primary sources of energy such as coal, oil, and natural gas to secondary sources such as fuel oil, gasoline, and electricity and includes losses incurred in the transmission of electricity from the power plant to its final destination. The magnitude of these losses is affected by the sophistication of technology used in the conversion processes as well as by how well plant and equipment are maintained. The significant drop in this ratio suggests that there has been a serious decline in energy efficiency, reflecting the deteriorating infrastructure in both the electricity and refining industries. According to Mesa-Lago, the national electricity system in June 2005 functioned at only 50 percent of capacity, with blackouts lasting seven to twelve hours on a daily basis.[20] These blackouts led the government to purchase diesel generators to supplement faltering power plants, but such generators are significantly less efficient than a properly functioning central generating plant.

Forecasting Future Energy Balance: Future Energy Demand

Energy forecasting is difficult in the best of cases. For Cuba, it is even more challenging, since future energy demand will be affected by policy changes that will emerge in the coming decades. It is hard to predict when a major change will occur and how it will play out inside Cuba. The only certainty is that the current economic development model has not been successful and will be modified or swept aside at some point. The experience of other countries that have shifted from a highly centralized command economy to one that is more decentralized and market oriented is marked by extremes. Countries in the former Soviet Union and eastern Europe suffered through a decade of economic turmoil during their transition to a market-based economy. On the other hand, China and Vietnam have experienced remarkably high rates of growth during their transitions. Cuba will be a special case and will not necessarily follow either of these patterns. Its proximity to the United States, the presence of a large expatriate Cuban community in which many members have close relatives in Cuba, and its unique history and culture will all play a role in its future direction.

In the absence of knowing how and when a transition will occur, our estimates of future Cuban energy demand are based on assumptions of various per capita income growth rates: 2 percent, 3 percent, and 5 percent per annum. These scenarios sidestep the issue of what Cuba's economic and political system will be in the future, especially during any "transition" phase. Clearly, if the experience of the former USSR is in store for Cuba, the rate of growth will be negative for a period of time and it could be a decade before there is modest growth. If China is the relevant model, then even our most optimistic scenario of 5 percent growth is conservative.

We derive future Cuban energy demand using two different exercises. First, we compare Cuba's energy use to that of countries that share some attributes with Cuba and ask what Cuba's demand would be if it had the same per capita energy use as those countries at a comparable level of per capita income. Second, we use an econometric model developed by Kenneth Medlock and Ronald Soligo that forecasts end-use-sector energy demand as a function of per capita income.[21]

Table 4-2 compares the structure and level of energy use of Cuba with that of several other Caribbean and Central American nations that share similar characteristics. For illustrative purposes, the authors have chosen the Dominican Republic, Costa Rica, and Jamaica—three countries that, like Cuba, have important tourist industries—and Guatemala, which is closer to Cuba in both population and per capita income.[22] Per capita GDP is presented in 2000 purchasing power parity dollars.[23]

Per capita total final consumption (TFC) of energy generally varies with per capita income. For Cuba and Guatemala, two countries with similar per capita income, TFC is approximately the same. For the richer countries, Costa Rica and Dominican Republic, TFC is higher. (Jamaica is an exception to this.)

Table 4-2. Breakdown of Energy Use, 2006[a]

	Cuba	Costa Rica	Dominican Republic	Jamaica	Guatemala
Population (millions)	11.3	4.4	9.6	2.7	13.0
Per capita income[b]	4,301	9,622	7,617	3,903	4,110
Per capita energy use (ktoe)					
Residential/commercial	110.5	203.5	239.9	177.9	315.6
Transportation	84.8	349.2	215.1	443.2	131.8
Industrial/agricultural/other	329.4	215.9	134.8	414.7	103.8
Total final consumption (TFC)	524.7	768.6	589.7	1,035.8	551.2
Total primary energy supply (TPES)	944.3	1,039.7	815.8	1,721.2	629.4
TFC/TPES	0.56	0.74	0.72	0.60	0.88
Energy source as percentage of total					
Coal and coal products	0.2	0.9	6.4	0.5	4.8
Crude oil, natural gas liquids, and feedstocks	52.6	15.0	26.0	22.3	1.7
Petroleum products	26.9	32.7	44.5	66.4	38.0
Natural gas	8.3	0.0	3.5	0.0	0.0
Hydro	0.1	12.4	1.5	0.3	4.0
Geothermal and solar	0.0	23.4	0.0	0.0	0.0
Combustible renewables and waste	11.9	15.5	18.0	10.5	51.5

Source: International Energy Agency, Energy Balances for Non-OECD Countries (www.iea.org), online subscription database.

a. The 2006 data are the latest available in the International Energy Agency database.

b. In 2000 PPP (purchasing power parity) dollars. This figure has recently been revised downward, to $3,500. See *CIA Factbook* (www.umsl.edu/services/govdocs/wofact2006/geos/cu.html#Econ).

The pattern of energy use, however, is very different among these countries. In Cuba, the extremely low energy use in the transportation sector is a consequence of severe restrictions imposed on the transportation sector in the wake of the cutoff of Soviet aid, as discussed earlier. Per capita energy consumption in the residential and commercial sectors is also low compared with other similar countries, a possible consequence of an underdeveloped commercial sector as well as a limited number of electrical appliances per capita. Frequent blackouts are another factor keeping consumption low.

Table 4-3 shows what the total primary energy supply would be in Cuba if its per capita level of final consumption were the same as that in the Dominican Republic and Costa Rica and at similar per capita income levels. Because of the large discrepancies in the transformation losses between Cuba and other countries and because it is unrealistic to think that Cuba can restore its energy infrastructure quickly, we assume that transformation losses as

Table 4-3. Three Energy-Use Forecasts for Cuba under Three Growth Rates—Comparison with Dominican Republic and Costa Rica

	2015			2020			2025		
Per capita growth rate	2%	3%	5%	2%	3%	5%	2%	3%	5%
Cuba's projected real GDP per capita[a]	$5,140	$5,612	$6,672	$5,675	$6,505	$8,515	$6,265	$7,542	$10,868
Energy-use comparison (ktoe)									
If like Dominican Republic[b]	9,950	11,312	12,142	10,708	11,702	NA[c]	11,127	10,439	NA
If like Costa Rica[b]	10,938	11,285	13,037	10,682	12,340	11,844	11,753	9,969	NA
Average	10,444	11,298	12,589	10,695	12,021	11,844	11,440	10,204	NA

Source: International Energy Agency, Energy Balances for Non-OECD Countries (www.iea.org), online subscription database.

a. Income in 2000 PPP (purchasing power parity) dollars.

b. Assuming transformation losses of 40 percent in 2015, 35 percent in 2020, and 30 percent in 2025.

c. NA means that per capita income has not yet reached those levels in Dominican Republic and Costa Rica.

measured by the ratio of TFC to TPES will improve from a recent five-year average of 42 percent to 40 percent in 2015, 35 percent in 2020, and 30 percent in 2025. That would roughly place Cuba in a position in 2025 where Costa Rica, the Dominican Republic, and Guatemala were in 2006. Clearly, if Cuba can move more quickly to modernize its electrical generating and distribution facilities, demand will be lower than our projections.

If Cuban per capita income grows at 5 percent per annum, its level in 2020 will be $8,515, roughly equal to the $8,506 per capita income of Costa Rica in 2003. If total per capita sectoral end-use of energy in Cuba in 2020 were to be the same as Costa Rica in 2003, total primary energy consumption in Cuba would rise from 10,639 ktoe in 2006 to 11,844 ktoe in 2020.

Alternatively, if Cuban per capita income grows at 3 percent per annum, it will reach $7,542 in 2025, comparable to the Dominican Republic in 2006, with per capita income at $7,617. Cuba's energy consumption in 2025, using the same per capita sectoral end-use as the Dominican Republic in 2006, would be 10,439 ktoe.

Our alternative approach to estimating energy demand is to use a model developed by Medlock and Soligo that projects total final consumption in each of the end-use sectors. The model assumes that the income and price elasticities of demand for each sector are the same for all countries, and we assume that those elasticities will also apply to Cuba. In that sense, we assume that Cuba will evolve into a country that looks recognizably like others in terms of the relationship between energy use and income.

Energy demand will grow much more slowly than the model predicts if Cuba continues to discourage private ownership of automobiles and energy use in households remains low, because of either housing shortages or the cost of acquiring energy-consuming durables such as air conditioning. On the other hand, if restrictions against private automobile ownership are lifted, we might find that demand for energy in the transport sector is underpredicted as Cubans rapidly acquire automobiles to satisfy pent-up demand. Similarly caveats apply to the tourist industry, which is energy-intensive, but has not grown very much in the last few years. Future growth will be affected significantly if, for example, the United States relaxes restrictions on travel.

Our estimates, shown in table 4-4, assume that energy prices will average $55 a barrel in 2010 and remain constant in real terms thereafter. Given recent volatility in prices and the levels prevailing as of spring 2010, this may be an optimistic scenario. Clearly energy demand will be less if prices rise.[24]

Table 4-4. Energy Demand Forecasts for Cuba under Three Growth Rates, by Sector

	2015			2020			2025		
Projected per capita growth rates (%)	2	3	5	2	3	5	2	3	5
Real GDP per capita (US$)	5,140	5,612	6,672	5,675	6,505	8,515	6,265	7,542	10,868
Sector (demand in kilotons of oil equivalent)									
Residential and commercial	1,126	1,161	1,230	1,157	1,231	1,376	1,230	1,353	1,597
Transportation	1,005	1,056	1,161	1,111	1,216	1,442	1,243	1,417	1,809
Industrial and other	4,236	4,351	4,565	4,543	4,744	5,090	4,853	5,130	5,554
Total final consumption	6,366	6,568	6,956	6,812	7,190	7,907	7,326	7,899	8,959
Total primary consumption[a]	10,610	10,480	11,593	10,947	11,062	12,165	10,465	11,285	12,799
TPEC in barrels per day[b]	183	181	200	189	191	210	181	195	221
TPEC per capita[a]	900	889	984	906	916	1,007	845	911	1,033
Total primary consumption[c]	10,976	11,744	12,631	11,324	12,397	13,620	11,993	13,633	15,447
TPEC per capita[c]	931	997	1,072	937	1,026	1,127	968	1,101	1,247

Source: Authors' compilation.

a. Transformation losses are assumed to be 40 percent in 2015, 35 percent in 2020, and 30 percent in 2025.

b. Assuming 6.3 barrels = 1 ton.

c. Assuming transformation losses remain at 2006 levels of 42 percent.

We also assume that population will grow at 0.5 percent per annum, a number that is very low by Latin American standards but is nonetheless higher than the actual growth rate in Cuba over the last decade of 0.28 percent. Finally, we calculate total primary energy consumption (TPEC) from total final (end-use) consumption (TFC) under the assumption that Cuba will gradually reduce energy transformation losses to 40 percent in 2015, 35 percent in 2020 and 30 percent in 2025. That would place Cuba in a position in 2025 roughly where Costa Rica, the Dominican Republic, and Guatemala were in 2006. We show a second set of calculations of energy demand if transformation losses are not reduced, to highlight the influence of transformation efficiency on energy demand. In this case, we assume that efficiency losses will continue into the future at 42 percent, the average level prevailing during the five-year period from 2002 to 2006.

The estimates from the two methods are reasonably consistent with one another. At a 3 percent rate of growth TPEC would be between 11.1 and 12.3 thousand ktoe by 2020 and between 10.2 and 11.3 thousand ktoe by 2025. Demand drops slightly, even though per capita income increases, because of assumed increases in efficiency. At a 5 percent growth rate, demand will be between 11.8 and 12.2 thousand ktoe by 2020 and 12.8 by 2025.

Taking the average of the high and low estimates for each case, we generate demand in terms of barrels of oil equivalent per day as follows: At a 3 percent growth rate demand will be 202,000 barrels of oil equivalent a day in 2020 and 207,000 by 2025. At a 5 percent growth rate demand would be 186,000 and 221,000 barrels a day, respectively. For comparison purposes, TPES in 2006 was 10,639 ktoe, the equivalent of 184,000 barrels a day. Future energy demand will increase by less than 25,000 barrels per day by 2025 at a per capita growth rate of 3 percent per annum and by about 37,000 barrels per day at a 5 percent growth rate. The assumption of reducing transformation energy losses is critical. If losses continue at the average level prevailing from 2002 to 2005, total demand in 2025 will be higher by 2,648 ktoe, or about 45,000 barrels per day, at 5 percent per capita growth.

Supply

As pointed out earlier, the U.S. Geological Survey has estimated that Cuba has mean "undiscovered" reserves of 4.6 billion barrels of conventional oil and 9.8 trillion cubic feet of gas in the North Cuba Basin. The USGS defines "undiscovered recoverable reserves (crude oil and natural gas)" as "those economic resources of crude oil and natural gas, yet undiscovered, that are estimated to exist in favorable geologic settings."[25] Recovery of these deposits is technically feasible, given current technology, but not necessarily economically feasible, since feasibility will depend crucially on oil prices as well as production costs. The USGS develops a probability distribution of these potential reserves. Its high estimate puts them at 9.3 billion barrels of oil and 21.8 trillion cubic feet of gas. Cupet claims the country has 20 billion barrels of recoverable oil in its offshore waters, and asserts that the higher estimate is based on new and better information about Cuba's geology than that reported by the USGS.[26]

The translation of estimated reserves into projected annual production flows is very difficult, and depends on a large number of factors. Of great importance is the issue of how much of the technically recoverable reserves are economically recoverable. This will be determined by future technological developments as well as prices of oil and the inputs required to develop and produce the oil. The flow rate will be determined in part by the nature of the geological structure of any find and the size and distribution of deposits within it. Absent knowledge of these variables, forecasts of supply are at best illustrative.

However, given our demand estimates, an average production of 200,000 barrels a day is all that Cuba needs to become self-sufficient in energy use under most of our assumptions. If reserve estimates were to be realized, this production level would be a very conservative estimate of possible production flows. In chapter 2 of this volume, Jorge Piñón states that Cuba's heavy oil production potential in onshore and coastal offshore areas *alone* could "grow to an amount in excess of 75,000 barrels per day" once Cuba gains access to technology and capital. If this forecast is borne out, Cuba would need only 125,000 barrels per day from offshore areas to be self-sufficient. Elsewhere, Piñón has suggested that Cuba could produce as much as 525,000 barrels per day when deepwater reserves are developed.[27]

In the case of gas, Peter Hartley and Kenneth Medlock have developed the Rice University World Gas Trade Model, which estimates production around the world on the basis of projections of demand and costs of production, and transport. The model forecasts that Cuba's natural gas production could rise rapidly and average more than 150 billion cubic feet per year.[28] On a barrel of oil equivalent basis, this would amount to roughly 77,000 barrels per day.

Cuba's Ethanol Potential

It is natural to associate Cuba with sugar. At one time, Cuba was the world's largest exporter of sugar. It was a major supplier to the United States before the Cuban Revolution, and to the Soviet Union in the 1970s and 1980s. But the industry has undergone a steep decline since major trade with the Soviet Union ended. Sugar production, as high as 8.1 million tons in 1988, had fallen to 1.25 million tons by 2009.[29] Acreage devoted to sugar was reduced by over 60 percent from 2002 to 2008. Sugar mills have been closed, with the number of plants falling from 156 to only 85. In 2006, output of raw sugar was approximately 1.2 million tons, reportedly the lowest output since 1908.[30]

Oddly enough, the retreat of the Cuban sugar industry has occurred at a time when many countries have been adopting policies to add ethanol, which can be made from sugarcane, to their transportation fuel portfolio. Despite the fact that Cuba is dependent on oil imports and is aware of the demonstrated success of Brazil in using ethanol to achieve energy self-sufficiency, it has not embarked on a policy to develop a larger ethanol industry from sugarcane.

In response to recent increases in ethanol prices, there is some support in Cuba for increasing ethanol production. A member of the Cuban Academy of Sciences, Conrado Moreno, has indicated that there are plans to upgrade eleven of the seventeen Cuban refineries to add annual production capacity of as much as 47 million gallons.[31] It remains to be seen whether this will happen without the support of top administration officials, especially Former President Fidel Castro.

Castro has rightly pointed out that there can be a direct trade-off between using land for food production and for ethanol. And in many areas of the world, the shift in land use to crops for ethanol has resulted in rapidly rising costs for food. There are also trade-offs between increasing acreage devoted to crops for ethanol and other objectives such as issues related to climate, environment, and biodiversity. In Brazil, for example, increasing acreage under sugarcane cultivation has resulted in shifting other crops to newly cleared areas, often in the rainforest, a process that ultimately could have devastating effects on climate and biodiversity within and beyond Brazil.

Cuba, however, has had a traditional comparative advantage in the production of sugar. Although some of the land used for sugar in the past is being shifted to food crops and reforestation, much of it is not currently being cultivated at all. Thus, for Cuba a restoration of the sugar economy does not necessarily have to involve environmental and food production trade-offs.

A Brief History of Sugar Cultivation in Cuba

Commodity markets are notoriously volatile and countries dependent on one or two commodity exports have always been subject to particularly harsh swings in economic activity. Cuba, traditionally dependent on sugar exports, has fitted into this pattern. Cuba's membership in the CMEA (Council for Mutual Economic Assistance), the trade area set up by the USSR for trade among Communist states, provided temporary stability for sugar prices in the 1980s, but ultimately the political changes in 1989 once again subjected Cuba to the vagaries of market fluctuations in sugar prices.

The heyday of the Cuban sugar economy was in the first three decades of the twentieth century, when sugar output increased fivefold, from less that 1 million tons in 1895 to 5.4 million tons in 1929.[32] Producing roughly 23 percent of the world's sugar (cane and beet sugar) and 36 percent of the world's cane output, Cuba was the world's largest producer and exporter of sugar.[33]

Cuban sugar production then declined to 2.1 million tons by 1932, in response to the worldwide depression and the imposition of the Smoot-Hawley tariffs in 1930 by the United States, Cuba's most important export market. Prices had already fallen by 38 percent between 1927 and 1929; by 1932 they had fallen an additional 57 percent, reaching the lowest level in the pre–World War II period.[34] Cuban sugar producers were further injured by the addition of U.S. sugar quotas in 1934 that favored producers in Hawaii, Puerto Rico, and the Philippines and limited Cuban exports to the United States to levels below those prevailing in the 1920s.[35]

During World War II, sugar production and exports recovered in tandem with the U.S. economy. After the war, international sugar agreements were implemented that helped stabilize prices by limiting world production. Cuban production came back to over 5.6 million tons by 1950, roughly where it was in the late 1920s, but exports to the United States were now governed by a system of tariffs and quotas. The dominant role of Cuba in international sugar markets and the inelastic demand for the product reduced the role of sugar as an engine of further growth and development in Cuba.

The postrevolutionary government chose to stress industrialization and agricultural diversification and to deemphasize sugar as its growth engine.[36] Alas, as many Latin American countries also discovered, an import substitution strategy requires ample supplies of capital and foreign exchange to finance investments in industry and the necessary complementary infrastructure, much of which has a large imported component. In the absence of private foreign investment, Cuba had to rely on its exports to provide the foreign exchange resources necessary for these investments. With sugar as its primary exportable commodity, Cuba's ambitions were quickly constrained by its loss of access to the U.S. sugar market once sanctions were imposed on the new regime.

Figure 4-1. Cuban Sugar Production, 1960 to 2010

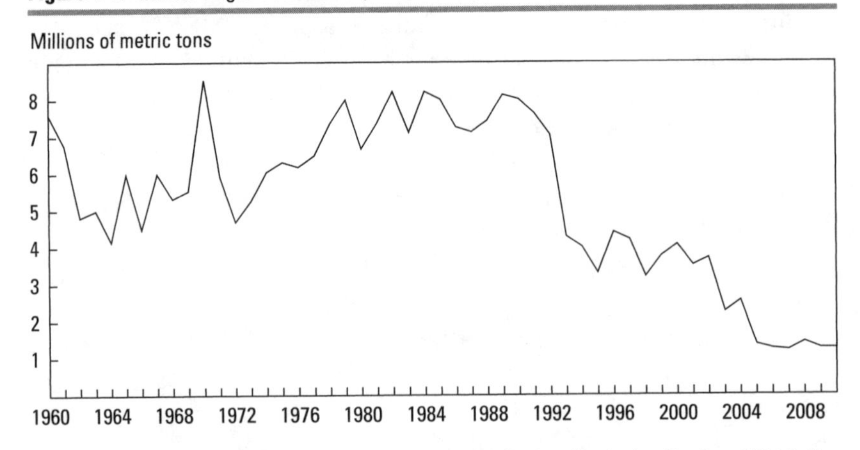

Source: U.S. Department of Agriculture, Foreign Agricultural Service, "Production, Supply and Distribution Online" (www.fas.usda.gov/psdonline/psdQuery.aspx).

Sugar resumed its dominant role in 1963, when Cuba entered into a trading relationship with the USSR and Eastern Europe whereby Cuba was to become the major sugar supplier to these countries. Cuba launched an ambitious plan to produce 10 million tons by 1970, but actual production fell far short of this target despite a focused effort by the government. Still, Cuba produced 8.5 million tons of sugar, the highest production level ever achieved.

When Cuba formally joined the Council for Mutual Economic Assistance (CMEA) in 1972, it did so under a generous arrangement with the USSR, which was prepared to pay Cuba a sugar price that was substantially higher than international market prices. The success of this arrangement for Cuba is apparent in the statistics for Cuban sugar production (see figure 4-1). Production grew rapidly and fluctuated between 7 million and 8 million tons during the 1980s.[37]

The collapse of the Soviet Union and the CMEA trading bloc was a disaster for the Cuban sugar industry. Once the lucrative communist bloc agreement that had provided stability was ended, Cuba was faced with global competition. However, years of high prices and the absence of competitive pressures had resulted in a loss of efficiency which, in combination with other problems faced by the agricultural sector in general, led to a period of continuous decline for the industry.

In 2002 the Cuban government launched a major restructuring of the sugar sector: sugarcane acreage was severely reduced and of the 156 sugar mills that existed at that time, 71 were closed permanently, 14 were devoted solely to the production of molasses and raw sugar for animal feed, and 64 were dismantled to be used as spare parts for the remaining mills.[38]

These reforms had a major impact on employment and resulted in the migration of farm labor to urban areas. Estimates of displaced workers range from 100,000 to 213,000 out of the 400,000 previously employed in the industry.[39] The policy's intention was to maintain sugar output at 4 million tons, but announced targets were not met. Like other areas of the economy, the sugar industry was plagued by aging plants and equipment and a lack of spare parts for maintenance. As of 2005, writes Carmelo Mesa-Lago, "only 70 percent of the mills operated and [these] were affected by frequent breakdowns."[40] In addition, the scarcity of foreign exchange limited imports of fertilizer and other supplies for the industry. Cuba actually had to import sugar in order to meet domestic demand and its export commitments to China. The situation prompted President Castro, reflecting frustration with the failure of the industry to meet targets, to say, "Sugar belongs to slavery times and will never come back to this country." Disillusionment with the sugar industry lies behind Castro's lack of support for the development of an ethanol industry.

Cuba's Future Sugar Industry: Ethanol Scenarios

The success of the Brazilian sugarcane and ethanol industry suggests that, despite former President Castro's views on the impossibility of restoring a viable Cuban sugar industry and the impact of sugar cultivation for ethanol production on food supplies, the Cuban sugar industry could have a promising future. The increasing use of biofuels in the transportation fuel mix in the United States and Europe provides a stable and growing market for ethanol, especially sugarcane-based ethanol, which is cheaper to produce than biofuels from other crops. The United States, under the Energy Independence and Security Act of 2007, increased the renewable fuels standard (RFS) to require that the use of biofuels gradually increase, to 36 billion gallons by 2022. Legislators intended that 16 billion of this consumption would come from cellulosic ethanol, but so far the development of a cost-effective production technology has been slow, leaving the market to corn- and sugar-based ethanol.

In 2009 the U.S. consumed 11.1 billion gallons of ethanol, almost all of it produced in the United States. U.S. policy favors domestic ethanol production by imposing an import tariff of 54 cents a gallon in addition to a

2.5 percent ad valorem tariff. Tariffs have limited ethanol imports into the United States, but higher prices in Europe have also been a factor. As of 2009, the United States has been suffering from an excess of production capacity, which has depressed prices in the States relative to other importing countries. But as higher U.S. renewable fuel targets kick in and U.S. prices recover from overinvestment in capacity, imported sugar-based ethanol will be competitive with higher-cost U.S. corn-based ethanol in coastal regions of the United States, even if U.S. tariffs persist. Given the high costs to transport corn-based ethanol to coastal regions from the U.S. Midwest by rail or truck.[41] Cuba's location gives it a large transport cost advantage over both domestic and foreign rivals.

Our analysis suggests that Cuba can produce 2 billion gallons of ethanol per year, equivalent to 94,500 barrels per day of gasoline, after adjusting for the differences in energy content. To arrive at this estimate we consider several factors that help determine ethanol output:

—The amount of land planted with sugarcane

—Yields (the amount of sugarcane harvested per hectare planted)

—The industrial yield (the amount of ethanol that can be produced from one ton of sugarcane)

—The proportion of sugarcane devoted to the production of sugar and other non-ethanol products

Amount of Land Planted with Sugarcane

Figure 4-2 shows the area of sugarcane harvested each year from 1961 to 2008. In 1970, the year of the ambitious campaign to produce 10 million tons of sugar, the area harvested was 1.5 million hectares, the highest level in the post–World War II period. Between 1971 and 1989 the area harvested averaged 1.28 million hectares, fluctuating between 1.14 million and 1.42 million hectares. After the collapse of the USSR and the end of Soviet aid, the harvested area plummeted, reflecting at first the decline in imported fuel, fertilizer, and other inputs and later, the decision to restructure the industry by shutting down inefficient sugar refineries and switching farms to pasture or other crops.

Since the special period in the early 1990s, Cuba has moved to diversify its agricultural sector in order to emphasize food security. It's not clear whether this was a response to economic and political conditions at the time or represents a permanent shift of agriculture away from depending so heavily on

Figure 4-2. Amount of Land Planted in Sugarcane, 1961 to 2007

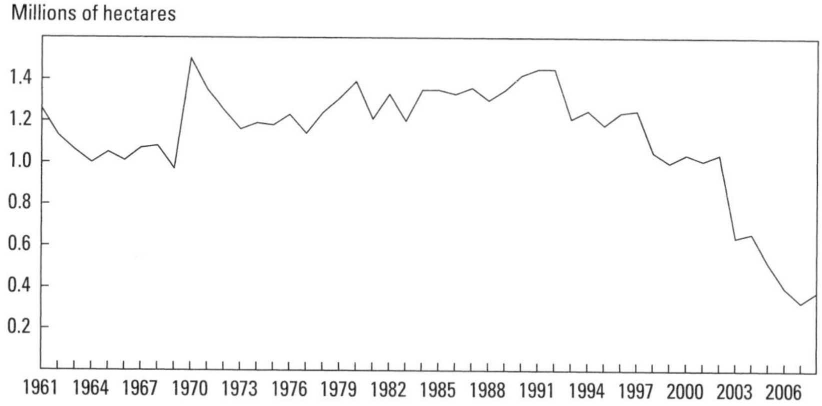

Millions of hectares

Source: Food and Agriculture Organization of the United Nations (http://faostat.fao.org/site/567/Desktop Default.aspx?PageID=567#ancor).

one crop. More recently, in 2008, the Cuban government announced grants of unused land to all private, cooperative, and state farms, as a spur to enhance domestic food production. The introduction of the plan was a response to the fact that in 2007, 55 percent of agricultural land remained idle, an increase from 46 percent in 2002.[42]

The shift in acreage devoted to food crops has not been successful in terms of increasing food output,[43] but reforms to give farmers more discretion in how they operate might produce better results in the future. But significantly increasing acreage devoted to food crops will not be easy. Food crops are much more fragile than sugarcane, requiring more labor, weeding, pest control, and oversight than cane, which has been referred to as the "widow's crop" because it requires relatively little attention. As noted previously, thousands of farm workers have migrated to urban areas and it will be difficult to lure them back. If economic sanctions are removed and Cuba enters the international commercial system, food security will be less important, and Cuban agriculture will be more likely to respond to international prices. Historically, Cuba has had a comparative advantage in producing sugar, not food crops; so opening the economy to freer trade might favor a return to the dominance of sugar and development of an ethanol industry. More recently, Cuba has expressed interest in producing and

Figure 4-3. Total Cuban Sugarcane Yields, 1961 to 2007

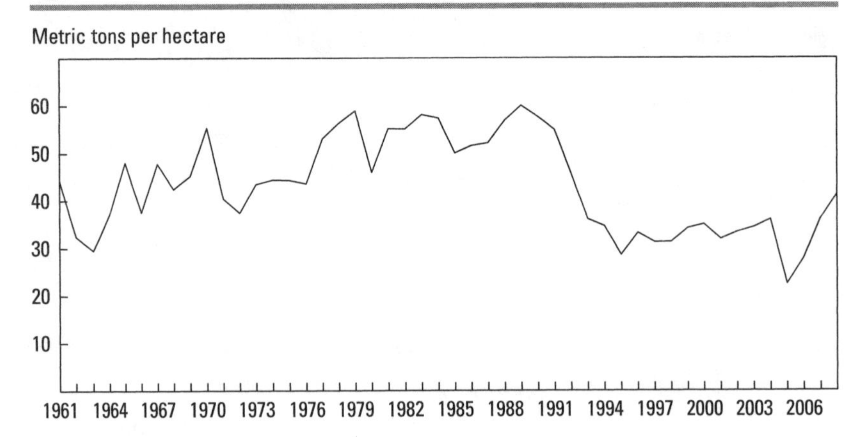

Source: Food and Agriculture Organization of the United Nations (http://faostat.fao.org/site/567/Desktop Default.aspx?PageID=567#ancor).

exporting soybeans, and the Brazilian government has offered "technical assistance and seed in order to grow soybeans on an industrial scale."[44] Soybeans have many uses, including as a feedstock for the production of biodiesel, but it is not clear at this point whether soybeans represent a more efficient use of Cuban land than sugarcane.

Sugarcane Yields

Sugarcane yields are highly variable—affected by weather conditions and other factors. Figure 4-3 shows sugarcane yields since 1961 and the decline in recent years as the industry has contracted. Yields that had fluctuated between fifty and sixty tons per hectare fell to twenty-eight in 2006.

Industrial Ethanol Output Levels

Table 4-5 shows the level of ethanol output per hectare of land that is devoted to the production of sugarcane targeted for ethanol production. Output in liters is shown as a function of sugarcane and distillery yields.

At a sugarcane yield of 75 tons per hectare and ethanol yield of 75 liters per ton (5,625 liters per hectare), an output of 7.6 billion liters, or 2 billion gallons, of ethanol requires approximately 1.33 million hectares of sugarcane. At 80 tons per hectare, it would require only 1.26 million hectares to produce 2 billion gallons. Finally, if Cuba achieves yields currently experienced in the Center-South region of Brazil of 84 tons per hectare and 82 liters per ton of

Table 4-5. Ethanol Output
Liters per hectare of sugarcane

		Sugarcane yield (metric tons per hectare)					
		32.3	55	75	80	85	90
Distillery yields (liters/ton)	70	2,261	3,850	5,250	5,600	5,950	6,300
	75	2,423	4,125	5,625	6,000	6,375	6,750
	80	2,584	4,400	6,000	6,400	6,800	7,200

Source: Authors' calculation.

cane (6,888 liters per hectare), it will need only 1.10 million hectares of sugarcane to achieve this volume.[45]

Sugar versus Ethanol

The amount of ethanol produced will also depend on how much of the sugarcane is used to produce sugar and other non-ethanol products. In 2009 Cuba produced 1.25 million metric tons of sugar on 380,000 hectares with very low yields of 41.3 tons per hectare. At an improved yield of 75 tons per hectare, 1.25 million tons of sugar would have required only 209,150 hectares, which at 5,625 liters of ethanol per hectare, would reduce ethanol output by 1,175,625 liters (310,000 gallons).

Sugar prices rose very quickly in 2009 to levels that are high by historical standards, approaching 25 cents a pound.[46] At these prices, producing and exporting sugar is more attractive than ethanol. But these prices are the temporary consequence of bad weather in other sugar-producing areas and will not be sustained. Both sugar and ethanol are commodities that will trade on the basis of price, and since entry into those industries is relatively unconstrained, competition will push prices down toward costs. When sanctions are lifted. Cuba will be able to benefit from the fact that it is an island economy with easy access to cheap marine transport—and the close proximity to the United States. Sugar imports in the United States are limited by quotas, so import volumes cannot change regardless of price. However, ethanol is protected by tariffs so imports can increase if domestic (U.S.) prices get too far ahead of world prices.

The fact that sugar exports are an alternative to ethanol is an additional argument for the development of an ethanol industry. To the extent that sugar and ethanol prices are not closely correlated, Cuba can alter its output mix between the two products to take advantage of variations in sugar and ethanol prices and thus smooth out fluctuations in export revenues as well as maximize the income from its sugarcane industry.

Issues in Achieving Cuba's Ethanol Potential

As noted, estimates of Cuba's ethanol potential will depend on assumptions about the amount of sugarcane that can be planted and harvested, as well as what sugarcane yields can be achieved. More ambitious assumptions will yield higher outputs. For example, Juan Sanchez assumes that Cuba could devote 2 million hectares to sugarcane with yields of 80 tons per hectare and 83.6 liters per ton (6,688 liters per hectare). He projects ethanol output at 13.4 billion liters, or 3.5 billion gallons.[47]

Three and a half billion gallons seems unrealistic for the foreseeable future. There is some question as to whether Cuba could ever again attain the 1.5 million hectares of sugarcane harvested in 1970, let alone 2 million. According to Brian Pollitt, the 1970 harvest was achieved only by cutting cane that would normally be left to mature for another season in order to produce a higher sugar yield in the following year.[48] Obviously this is not a sustainable practice if optimal yields are to be achieved.

Two billion gallons can be produced with a harvested area of 1.33 million hectares and a yield of seventy-five tons per hectare. That area of cultivation is not too far from the average harvest of 1.28 million hectares that Cuba was able to maintain during the 1970s and 1980s. Yet reaching 1.33 million hectares will require time and substantial investment in farm machinery and restoration of the land, which has been neglected and compacted by the use of heavy Soviet-built harvesting machinery. The land will also have to be tilled and newly planted with sugarcane.

Achieving higher sugarcane yields will also require time and investments to acquire or develop higher-yielding sugarcane varieties. Cuban yields averaged only fifty-eight tons per hectare during the 1970s and 1980s, substantially below the seventy-five tons per hectare needed to produce 2 billion gallons of ethanol. Yet other countries, as noted, have achieved or exceeded that yield, and some private Cuban farmers are reported to have achieved even higher yields of 100 tons per acre.[49] Yields, of course, are a function of other factors besides cane variety. The condition of the land, access to water and fertilizer, and other inputs would all need to be considered.

Finally, Cuba will have to undertake significant investments in distilleries, transport, storage, and distribution infrastructure if it wants to produce the levels of ethanol that the authors believe are achievable. Investment costs for the biorefineries alone will come to billions of dollars. For example, in 2006, corn-based ethanol plants in the United States cost roughly $1.88 per gallon for a capacity of 48 million gallons per year, and $1.50 per gallon for capacity of 120 million gallons per year (reflecting significant economies of scale). So even if all new plants in Cuba were built with the larger capacity, it would require $3 billion dollars (at 2006 prices) to build sufficient capacity to produce 2 billion gallons.

Looking at the Brazilian experience with ethanol is instructive. There, in 1975, the government introduced Proálcool, its national ethanol-production program, as a response to the oil shocks of that decade. It took several decades for Brazil to achieve current agricultural yields and ethanol output. Its approach to promoting ethanol use was to mandate that gasoline be mixed with 10 percent ethanol starting in 1975 and that this proportion should be increased to 25 percent by 1980. The government provided loans for the construction of ethanol plants and guaranteed the price of ethanol. Following the second spike in the price of oil in 1980, the government required Petrobras, the state-owned oil company, to supply ethanol to filling stations and offered a subsidy to auto firms to produce cars that could run on pure ethanol. It is estimated that the government spent over $16 billion (in 2005 dollars) between 1979 and the 1990s on subsidies and price supports to promote the industry.[50]

Cuba has the advantage of being able to learn from the Brazilian experience, even though the evolution of the industry in Cuba will certainly differ from that in Brazil. Exports could play a larger role at an earlier phase in the development of the Cuban industry. Domestic absorption of ethanol within Cuba will be constrained by the longevity of the existing vehicle stock (which burns only gasoline), the speed with which the number of motor vehicles is increased, and the extent to which new vehicles are "flexfuel" vehicles, able to burn both fossil and biofuels. Cuba has the additional advantage of a more robust international demand for ethanol than was the case when Brazil initiated its policy thirty years ago.

Cuba has opened the door to foreign investment in the ethanol sector, and Brazil has expressed interest in sharing its expertise in order to promote the use of ethanol and the development of a market where ethanol is traded like other commodities.

Ethanol and the Production of Electricity

The economics of ethanol production from sugarcane is enhanced by using the sugarcane waste (bagasse) to produce electricity by burning it. One estimate is that Cuban mills produce 20 and 40 kilowatt-hours per ton of sugarcane, depending on the age and efficiency of the steam turbines.[51] This is below the 55 kilowatt-hours reported for plants in Central America and significantly below the 100 kilowatt-hours per ton achieved by some Hawaiian mills.[52] Although bagasse is available only during the harvest season, these plants can be fueled with woodchips and other waste in at least part of the non-harvest season. Even at the modest yield of 55 tons of sugarcane per hectare and 55 kilowatt-hours per ton, a million hectares of sugarcane will produce roughly 3 billion kilowatt-hours of electricity, almost 20 percent of the 16.5 billion kilowatt-hours produced in Cuba in 2006. With higher yields, 1.3 million hectares could produce 4 billion to 5 billion kilowatt-hours.

The Structure of an Ethanol Industry

If Cuba decides to develop an ethanol industry it will have to decide on how to structure it. In particular, it will have to decide on the relative roles of the Cuban state and private citizens as well as the role of foreign companies.

There are several models that Cuba can choose from. One is to resuscitate a national, state-owned sugar industry with the addition of state-owned biorefineries. Sugarcane would be grown on state farms and cooperatives, processed in state-owned biorefineries, and marketed by an agency of the government. Past experience suggests that the state has not been able to operate the sugar industry in a cost-competitive way. Recent land reforms are motivated by that experience. Agriculture depends on rapid decisionmaking in response to changing location-specific information such as weather patterns, soil conditions, and pest infestations. Successful agriculture depends on decentralized decisionmaking with proper incentives given to the decentralized manager, a lesson learned in all highly centralized economies. In addition to these efficiency considerations, the Cuban government would have great difficulty in raising the enormous amounts of capital necessary to revive large-scale sugar cultivation and construct biorefineries and other needed infrastructure if these were to be solely within the state sector.

Another option is to follow the policies used in the oil and nickel industries, where foreign private firms currently operate. These firms provide the technology, management expertise, and capital, while the state provides labor. Workers would have to be well paid and well treated—otherwise this approach might be politically difficult, since it would hark back to the sugar plantations of the prerevolution years. Under this model Cuba is able to get access to needed resources, yet still maintain "control" of the industry and the egalitarian income policies that characterize the Cuban socialist model.

Finally, Cuba can continue its agricultural reforms and encourage sugarcane cultivation by individual farmers or cooperatives who could sell their output to biorefineries owned and operated by privately owned domestic or foreign firms. This option might attract foreign capital and expertise in the biorefinery end of the industry, but it is difficult to see where private and cooperative farms would get access to the large amount of capital needed to rebuild the agricultural capacity of the country. Farmers would require access to credit to purchase inputs needed in the cultivation of sugarcane. In the absence of U.S. sanctions, Cuba would have access to the resources from the international banking institutions (World Bank and the Inter-American Development Bank), but resources from these institutions come with controls and constraints that the Cuban government would find uncomfortable. Furthermore, relying on more independent farmers would also create a class of private and cooperative farmers whose incomes would not be subject to state control, and could lead to income inequalities.

Conclusion

Our intention in this chapter was to present the case that Cuba's energy potential is sufficient for Cuba to shift from its status as a net importer of roughly 100,000 barrels of oil a day to one of a net energy exporter. We have derived what we feel are conservative estimates of future energy demand and suggest that Cuba's oil production potential alone could probably satisfy future energy demand growth, provided that Cuba begins to do something about its abnormally high energy transformation losses. In addition, we suggest that Cuba could produce upwards of 150 billion cubic feet per day of natural gas, equivalent to 77,000 barrels of oil equivalent per day. Finally, ethanol production of 2 billion gallons per year could replace 94,500 barrels per day of gasoline as well as 3,000 gigawatt-hours of electricity—18 percent of current Cuban production—through cogeneration.

It is not possible to generate estimates of Cuban demand for specific fuels, since Cuba will have a choice of which to use domestically and which to export, depending on the relative prices of various fuels in international markets. But it is clear that Cuba has the potential of being a significant exporter of several energy resources, shifting the country from a nation where energy poverty has negatively affected overall economic performance to a country where energy surpluses could support economic growth.

The development of its energy resources could have a profound impact on Cuba's economy. Simply replacing current oil imports would release foreign exchange for other developmental uses. For example, at $60 a barrel, 100,000 barrels per day of imports has a market value of $2.2 billion a year, roughly equivalent to all the earnings from the tourist industry.[53] Energy exports will add a further significant boost to the Cuban economy. The experience of Brazil is instructive. In the 1970s Brazil found itself facing financial crises when oil prices spiked as a result of Middle East instability. By contrast, Brazil in 2007 and 2008—by then a net exporter of energy—saw less economic hardship arising from the dramatic increase in oil prices than other industrialized countries in those years.

Whether the scenarios discussed in this chapter are realistic can be established only when serious oil and natural gas exploration and development of Cuban assets begins. Cuba's nascent potential in ethanol also remains theoretical so far. However, the recent political transition in Cuba and the change in administration in the United States make this an ideal time to reevaluate U.S.-Cuba policy, taking into consideration humanitarian issues as well as energy potential. Having an additional supplier of energy to the U.S. market from only a few miles off shore can only contribute to the United States' energy security.

Notes

1. Amy Myers Jaffe and Ronald Soligo, "Potential Growth for U.S. Energy in Cuba," *Cuba in Transition* 12(2002): 422–30.

2. United States Geological Survey, "Assessment of Undiscovered Oil and Gas Resources of the North Cuba Basin, Cuba, 2004" (http://pubs.usgs.gov/fs/2005/3009/pdf/fs2005_3009.pdf).

3. David R. Mares and Nelson Altamirano, "Venezuela's PDVSA and World Energy Markets: Corporate Strategies and Political Factors Determining Its Behavior and Influence," *The Changing Role of National Oil Companies in International Energy Markets* (Houston: Rice University, James A. Baker III Institute and Japan Petroleum Energy Center, 2007).

4. Carmelo Mesa-Lago, "The Cuban Economy in 2006–2007," *Cuba in Transition* 17 (2008): 1–20.

5. Central Intelligence Agency, *The World Factbook,* "Cuba" (www.umsl.edu/services/govdocs/wofact2006/geos/cu.html#Econ; www.umsl.edu/services/govdocs/wofact2007/geos/cu.html#Econ; www.cia.gov/library/publications/the-world-factbook/geos/cu.html).

6. Measuring growth in the Cuban economy is contentious and some experts would dispute the IEA and CIA data on per capita GDP and GDP growth. See, for example, Camelo Mesa-Lago, "The Cuban Economy in 2004–2005," *Cuba in Transition* 15 (2006): 1–18; Mesa-Lago, "Cuban Economy in 2006–2007."

7. "Watching the World: Brazil Courts Cuba," *Oil and Gas Journal* 106, no. 3 (January 21, 2008): 41.

8. United Nations Data, Commodity Trade Statistics Database, United Nations Statistics Division (http://data.un.org/Data.aspx?q=null&d=ComTrade&f=_l1Code%3a76). Set filter on left for "Cuba" and desired year. See displayed data for nickel oxide sinters and intermediate nickel products.

9. Mesa-Lago, "Cuban Economy in 2006–2007."

10. See "Mariel Port Development Takes Off, Courtesy of Lula's Brazil" (http://store.businessmonitor.com/article/330331).

11. Patricia Grogg, "Cuba-China: Firm Friends and Excellent Business Partners," Inter-Press Service, Havana, February 20, 2006.

12. "Cuba and Russia Agreed on Joint Energy Project," Power-Gen Worldwide website, February 4, 2009 (http://pepei.pennnet.com/Articles/Article_Display.cfm?Section=ARTCL&SubSection=Display&PUBLICATION_ID=6&ARTICLE_ID=352359).

13. "Oil and Gas—Cuba," Sherritt.com (www.sherritt.com/doc08/subsection.php?submenuid=operations&category=operations/oil_and_gas_cuba).

14. Sherritt, *Annual Report 2008* (www.sherritt.com/doc08/files/financials/2008 Annual Report/Sherritt_AR08_full.pdf, pp. 7, 25).

15. Eric Watkins, "Cuba's Oil, Gas Production Rising, Politburo Member Says," *Oil and Gas Journal* 105, no. 35 (September 17, 2007): 49.

16. U.S.-Cuba Cooperative Security Project, "Cuba & Energy: A News Chronology," World Security Institute website (www.wsicubaproject.org/cubanenergy_052506.cfm).

17. Jonathan Benjamin-Alvarado, "Commentary on 'Cuba's Energy Challenge: A Second Look,' by Piñón Cervera," *Cuba in Transition* 15 (2006): 124 (http://lanic.utexas.edu/project/asce/pdfs/volume15/pdfs/benjaminalvarado.pdf).

18. International Energy Agency, Energy Balance Spreadsheet (www.iea.org); they may differ from other estimates. The term "energy balances" is standard usage (by the IEA and others) to depict data on energy supply and demand and their components—domestic production, imports and exports, and consumption.

19. Marcus Enoch and others, "The Effect of Economic Restrictions on Transport Practices in Cuba," *Transport Policy* 11, no. 1 (January 2004): 67–76.

20. Mesa-Lago, "Cuban Economy in 2004–2005."

21. Kenneth Medlock III and Ronald Soligo, "Economic Development and End-Use Energy Demand," *Energy Journal* 22, no. 2 (April 1, 2001): 77–105.

22. The Cuban per capita income statistic, $8,172, given in the IEA database (subscription database) is unrealistically high. Consequently, the authors have used the number $4,100, given in Central Intelligence Agency, *World Factbook* 2006 (www.umsl.edu/services/govdocs/wofact2006/geos/cu.html).

23. There is no consensus on the level of Cuba's per capita income. The per capita income in 2000 PPP (purchasing power parity) dollars given in the most recent IEA data (subscription online database) is roughly twice that reported in earlier versions based on 1995 dollars. The difference in the two series cannot be accounted for by

inflation in that short period. Mesa-Lago ("Cuban Economy In 2006–2007") has shown that when Cuba changed the base year used to estimate GDP in constant dollars from 1981 to 1997, reported GDP increased by 56 percent on average for each year from 1989 to 2000 and per capita income by 85 percent for 2001! In addition, Cuba introduced a new national accounting methodology in 2003 that had the effect of further increasing GDP. For our purposes the precise measurement of per capita income is not critical.

24. For example, the oil price forecast of the U.S. Department of Energy's Energy Information Administration (EIA) in its 2009 Annual Energy Outlook, "World Oil Prices in Three Price Cases 1980–2030" (www.eia.doe.gov/oiaf/ieo/graphic_data_world.html), assumes that prices, in its reference case, will rise in real terms to $121 per barrel by 2025, and total primary energy demand will only be 10,425 ktoe in 2020 and 10,531 ktoe in 2025 at a 5 percent growth rate in per capita income. This is in contrast to 12,165 and 12,799, respectively, with $55 per barrel oil, figures arrived at by means of simulations done with the Medlock-Soligo model discussed in the text.

25. U.S. Energy Information Administration, "Glossary—U" (www.eia.doe.gov/glossary/glossary_u.htm).

26. Jeff Franks, "Cuba Says May Have 20 Billion Barrels of Oil Offshore," Reuters, October 16, 2008.

27. For comparison, Equatorial Guinea currently produces approximately 400,000 barrels per day with proven reserves of 1.2 billion barrels in deep water offshore. If Cuba's offshore resources are comparable, 525,000 barrels per day is not an unreasonable estimate.

28. Production estimates were provided by the authors of the model. For a discussion of the model see Kenneth Medlock and Peter Hartley, "Rice University World Gas Trade Model," CEC Workshop Presentation, June 16, 2009 (www.energy.ca.gov/2009_energypolicy/documents/2009-0616_workshop/presentations/07_Medlock_The_Rice_World_Gas_Trade_Model.pdf).

29. U.S. Department of Agriculture, Foreign Agricultural Service, Downloadable Data Sets (www.fas.usda.gov/psdonline/psdDownload.aspx).

30. Marc Frank, "Cuba Plans Big Increase in Ethanol Production," Planet Ark website, June 21, 2006 (www.planetark.com/dailynewsstory.cfm/newsid/36927/newsDate/21-Jun-2006/story.htm).

31. "Cuba to Modernize Its Ethanol Production," Oryza News (website on rice), May 23, 2007 (http://oryza.com/Global-Rice/Bio-Tech-News/Cuba-ethanol.html).

32. Brian Pollitt has provided a detailed history of the Cuban sugar industry: Brian H. Pollitt, "The Cuban Sugar Economy: Collapse, Reform and Prospects for Recovery," *Journal of Latin American Studies* 19 (1997): 171–210. See also Pollitt, "The Rise and Fall of the Cuban Sugar Economy," *Journal of Latin American Studies* 36 (2004): 319–48.

33. Pollitt, "Rise and Fall of Cuban Sugar Economy," 320.

34. Pollitt, "Cuban Sugar Economy," 172.

35. Pollitt, "Rise and Fall of Cuban Sugar Economy," 321–22.

36. See Jorge F. Pérez-López and José Alvarez, "The Cuban Sugar Agroindustry at the End of the 1990s," in *Reinventing the Cuban Sugar Agroindustry,* edited by Jorge F. Pérez-Lopez and José Alvarez (Lanham, Md.: Lexington Books, 2005), p. 27.

37. Production, acreage, and yield data: Food and Agriculture Organization of the United Nations (http://faostat.fao.org/site/567/DesktopDefault.aspx?PageID=567#ancor).

38. Jorge F. Lopez, "The Cuban Economy in 2002–2003," *Cuba in Transition* 12 (2003): 507–21.

39. Ibid., 9.

40. Mesa-Lago, "Cuban Economy in 2004–2005," 7.

41. The Energy Forum of the James A. Baker III Institute for Public Policy and Rice University's Department of Civil and Environmental Engineering, "Fundamentals of a Sustainable U.S. Biofuels Policy," January 2010, p. 49 (www.bakerinstitute.org/publications/EF-pub-BioFuelsWhitePaper-010510.pdf).

42. Marc Lacey, "Cuba to Grant Private Farmers Access to Land," *New York Times,* July 19, 2008 (www.nytimes.com/2008/07/19/world/americas/19cuba.html).

43. Walfrido Alonso-Pippo and others, "Sugarcane Energy Use: The Cuban Case," *Energy Policy* 36 (2008): 2163–81.

44. Estaban Israel, "Brazil to Help Cuba Grow Soy on Industrial Scale," Reuters UK, May 30, 2008 (http://uk.reuters.com/article/idUKN3042480220080530).

45. Datagro (Brazil) Private communication with the authors.

46. Prices have continued to rise in 2010. See Index Mundi website, "Sugar Monthly Price" (www.indexmundi.com/commodities/?commodity=sugar).

47. Juan Tomás Sanchez, "Cuba y el Etanol: Proyeciones para una economía privada" [Cuba and ethanol: Projections for a private economy], *Cuba in Transition* 17 (2007): 199–205.

48. Pollitt, "Cuban Sugar Economy," 185.

49. Walfrido Alonso-Pippo and others, "Sugarcane Energy Use," 2168.

50. Estimate made by Datagro, reported by David Luhnow and Geraldo Samor, "As Brazil Fills Up on Ethanol, It Weans off Energy Imports," *Wall Street Journal,* January 16, 2006, p. A1.

51. These figures are for net export to the grid. See Roger Lippman and others, "Renewable Energy Development in Cuba: Sustainability Responds to Economic Crisis," April 1997 (http://tlent.home.igc.org/renewable%20energy%20in%20cuba.html).

52. Ibid.

53. Although Cuba is not currently paying market prices for its imports from Venezuela, it cannot assume that this favorable arrangement is permanent. Rather, Cuba should view Venezuelan aid as a bridge to the time when market prices will have to be paid.

five

Prospects for U.S.-Cuban Energy Engagement: Findings and Recommendations

JONATHAN BENJAMIN-ALVARADO

At the outset of this book we asked the question, "What would an ideal strategic energy policy look like for the United States, or any other country, for that matter?"

The Current Outlook

Mahmoud Amin El-Gamal and Amy Myers Jaffe partially answered that question for us in their paper "Energy, Financial Contagion and the Dollar,"[1] by setting out a detailed analysis of the tasks that a strategic energy policy should accomplish, among them:

—To ensure that markets operate efficiently so as to develop the infrastructure necessary to meet growing energy requirements

—To ensure the well-being of the human habitat and ecosystem

—To guarantee that mechanisms are in place for warding off and, if necessary, managing disruptions to energy supply

They provided a refined definition of what energy security is and how it is related to U.S. geostrategic interests specific to the discussion of Cuba, the Caribbean, and Latin America. Energy security is a critical consideration for three reasons.

1. U.S. energy independence is not attainable.
2. The policy instruments available to deal with energy supply disruptions are increasingly inadequate.

3. The United States needs to articulate a new vision of how to effectively manage international energy interdependence.

Where Cuba is a factor in that discussion, we were compelled to ask the following questions concerning the implications of a growing presence of these external actors in Latin American energy markets:

—In what way will the ongoing development and evolution of Cupet, Cuba's state oil company, limit or obstruct U.S. efforts to meet its strategic objectives? Any answers must account for the relationship among Cupet, PDVSA (Petróleos de Venezuela), and the Venezuelan state.

—What role can international oil companies play in the development of energy resources and infrastructure in Cuba in both the near and long term? Cuba is seeking to develop a production capability of its North Coast Reserves, leading national oil companies (NOCs) from nine different countries to sign lease agreements with the Cuban regime for offshore tracts.

—What impact will competition for scarce petroleum resources with Brazil, Russia, India, and China have on U.S. energy security, especially in light of the recent energy development agreement between Russia and Cuba, and China's advance into Latin American energy markets?

New players in Cuba and Latin America increase uncertainty over energy sources on which the U.S relies for its economic lifeblood. Many of these and other energy sources are controlled by NOCs:

—Seventy-seven percent of proven oil reserves globally are held by NOCs.

—Eleven percent of proven oil reserves are held by NOCs with equity access.

—Only 11 percent of proven oil reserves are open to international oil companies (IOCs); many of these are based in the United States.

The authors of chapter 2, Jorge R. Piñón and Jonathan Benjamin-Alvarado, find that there are a number of key issues to consider regarding the productive capacity of Cuba's oil and gas resources. First, Cuba has seen close to $2 billion of direct foreign investment since 1991 in its upstream oil and natural gas sector, with very good results. Crude oil liquids production reached a peak level of 65,531 barrels per day in 2003, up from 9,090 barrels per day in 1991. Since 2005 Cuba has seen its crude oil production level off at around 52,000 barrels per day. Second, Cuba's realized crude oil value could improve substantially once the country is able to monetize its heavy oil production by means of its own future heavy oil conversion refinery processing capacity, or to market its crude oil to U.S. Gulf Coast refining companies. Third, Cuba's onshore and coastal heavy oil production seems to have reached a plateau at around 52,000 barrels per day, but once Cupet has access to the services, technology, equipment, and capital available through independent U.S. oil and oil services and equipment companies (when the trade embargo is lifted or modified), Cuba's heavy oil production potential could grow to an amount in excess of 75,000 barrels a day.

Deficiencies in Cuba's oil-refining sector—including outdated technology that is unable to process heavy crude—coupled with an environmentally sensitive tourist industry will force Cuba to consider developing an energy policy that relies heavily on clean-burning natural gas as its fuel of choice for power generation. Cuba's future natural gas needs could be sourced as liquefied natural gas (LNG) from Trinidad and Tobago, as Puerto Rico and the Dominican Republic currently do, or from future Venezuelan production. A regasification facility to receive Venezuela-sourced liquid natural gas is being planned for the southern coast port city of Cienfuegos by Venezuela's PDVSA and Cupet. Two one-million-ton regasification trains are planned for 2012 at a cost of over $400 million. The natural gas is destined as fuel for that city's thermoelectric power plant, local industry, and future petrochemical plants.

As of late 2009, Cupet has consigned eighteen of the fifty-nine deepwater blocks in Cuba's Exclusive Economic Zone (EEZ) to seven international oil companies. Piñón and Benjamin-Alvarado caution against getting too excited about Cuba's immediate offshore oil potential—there are a number of obstacles to be overcome—yet the outlook is basically positive. Deepwater exploration is expensive and carries a high degree of geological and technical risk, risks that companies such as Repsol-YPF, Statoil–Norsk Hydro, and Petrobras certainly have the necessary deepwater expertise to handle. The price of crude oil would have to be over $65 per barrel in order to be worthwhile for most international oil companies to undertake today, and current prices are close to this.[2] If successful, it could take two to three years to bring the North Coast Basin deepwater project into full development, at an estimated total cost of $1 billion to $3 billion.

Future challenges in the upstream oil and gas sector need to be understood in terms of current and reported future international oil companies that are involved in Cuba's deepwater search for oil and gas: their competency, strategic objectives, and possible long-term contribution to the island's goal of becoming energy-independent. As long as the U.S. government's current economic and trade restrictions imposed on the government of Cuba remain in place, all companies, regardless of their nationality or technical competence, will have a very difficult time monetizing any newly discovered hydrocarbon resources, because they need access to the U.S. oil services and equipment sector.

In chapter 3, Juan Belt argues that Cuba has succeeded in extending electricity services to a large proportion of the population—a noteworthy achievement, accomplished with little regard for economic, financial, or environmental considerations. Power generation from burning liquid fuels, the predominant type in Cuba, is extremely costly when all the opportunity costs are taken into account. Additionally, Cuba's state-owned electric utility, Unión Eléctrica, experiences higher transformation losses of power and has lower labor productivity than other countries. Venezuela came to the rescue when Cuba lost fuel subsidies from the Soviet Union, but Venezuela's own fiscal accounts are now under significant pressure as a result of the sharp reduction in the price of crude oil. If the crude oil market remains depressed, Venezuela may have no option but to reduce or end its support for Cuba, which would jeopardize the island's precarious energy security situation. When we modeled Cuba's energy sector, we assumed that Venezuela would terminate its subsidies to Cuba by 2010.

The analysis uses the MARKAL/TIMES model, which facilitated a low-cost expansion plan for Cuba's power system. Unfortunately, Cuba does not have great potential for renewables, but together, wind power, photovoltaic cells, small hydropower, and bagasse represent a larger renewables potential than is currently realized.[3] Under all scenarios, domestic or imported natural gas will become the predominant fuel. Combined-cycle gas turbines would provide cheaper and cleaner power but would require an investment of $2.5 billion, and if the gas is imported, an additional investment will be needed to build a regasification facility. The Cuban government has announced plans to build a gas regasification facility in Cienfuegos; the estimated completion date is 2013.

Attracting such high levels of foreign direct investment, particularly during the current global economic crisis, will require significant reforms designed to reassure potential investors:

—Passing an electricity-sector law and establishing a public utility commission

—Modeling the energy sector to better determine the least-cost expansion path, including a more thorough analysis of the prospects for renewable energy

—Modifying tariffs gradually to reach full cost recovery, restructuring Unión Eléctrica through unbundling and corporatizing, and promoting independent power producer arrangements

—Developing operating contracts or concessions for existing assets

The United States should support these efforts. Initial funding possibilities include technical assistance for modeling the sector to determine options for improving efficiency and environmental sustainability and training government officials in the economic regulation of utilities. Modeling and training will enhance the skills of Cuban professionals and foster a dialogue with their peers in the United States. More cooperation between Cuba and the United States will enhance energy security in both countries.

In chapter 4, on Cuba's energy balance and potential for biofuels, Amy Myers Jaffe and Ronald Soligo echo Piñón's opinion that the potential for exploiting oil and gas reserves is indeed significant and that Cuba has large land areas that once produced sugar but now lie idle. These could be revived to provide a basis for a world-class ethanol industry. They estimate that Cuba could very easily produce 1 billion gallons per year of sugar-based ethanol. They make the case that Cuba has the potential not only to be self-sufficient in energy but also to become a net exporter. By 2000 the Cuban economy had made progress in recovering after the collapse of the Soviet Union and the ensuing special period. Total energy and crude supplies increased. Domestic production of crude more than doubled, and imports began to rise. Domestic production of gas continued to increase. The balance between the supply of crude and petroleum products shifted back toward its historical ratio favoring crude oil and oil consumption in all end-use sectors. These trends continued into 2009.

One troubling aspect of Cuba's energy profile is the ratio of total final consumption (TFC) to total primary energy supply (TPES), which is a measure of the energy lost during conversion—primarily in the generation of electricity but also in the refining process. It has shown a steady decline since 1990, with losses in transmission and delivery reducing the available supply of electricity from about 85 percent of all electricity produced to 71 percent in 2009. The significant drop suggests that there has been a serious decline in energy efficiency, no doubt reflecting the deteriorating electricity and refining industry infrastructure.[4] Forecasting Cuba's future energy balance is risky, since future energy demand will be affected by policy changes that are likely to emerge in the coming decades. It is hard to predict when a major change will occur and how it will unfold inside Cuba, but many experts expect economic policy changes to emerge, especially if the United States lifts sanctions. As Jaffe and Soligo state, "The only certainty is that the current model has not been successful and will be modified or swept aside at some point."

Should Cuba successfully tap its energy production potential, Jaffe and Soligo estimate that Cuba could produce as much as 2 billion gallons or

130,000 barrels per day of ethanol. Adjusting for the energy content of ethanol, this is the equivalent of 97,500 barrels per day of gasoline. Having three different energy streams will give Cuba the flexibility to exploit world price differentials and to choose which fuel to reserve for domestic production and which to export. In many areas of the world, the shift in land use to crops for ethanol has resulted in rapidly rising costs for food, but this is not the case in Cuba, which has had a traditional comparative advantage in the production of sugar because of its year-round growing season. Although some of the land used for sugar in the past is being shifted to food crops and reforestation, much of it is idle. Thus, for Cuba a restoration of the sugar economy does not necessarily have to involve the sort of trade-offs in food production and environmental quality that are issues in the United States and Brazil.

Cuba's ethanol potential is second to that of Brazil in Latin America. Of course, achieving high levels of ethanol production capacity in Cuba will take time. There are many obstacles to achieving an ethanol industry that could produce as much as 2 million gallons of ethanol output. Increasing the area under sugarcane will require substantial investment. The land has been neglected and much of it has suffered from compaction with the use of very heavy Soviet-built harvesting machinery. As previously mentioned, the land will have to be tilled and newly planted with sugarcane. Harvesting machinery has not been maintained and much of it will have to be replaced.

In addition, many sugar mills have been closed, and those that remain have not been properly maintained. Moreover, Cuba will have to undertake significant investments in distilleries and transport, storage, and distribution infrastructure if it is to produce the levels of ethanol that we believe are achievable. Investment costs for the biorefineries alone will come to billions of dollars. Cuba has the advantage of learning from Brazilian experience. The evolution of the industry in Cuba will differ from that in Brazil in that exports are likely to play a larger role at an earlier time period in the development of the industry. This is because there is a much more robust international demand for ethanol today than was the case when Brazil initiated its efforts.

Jaffe and Soligo conclude by stating that the next decade could be one in which Cuba becomes self-sufficient in energy and most likely becomes a net exporter. Cuban energy use may grow to as much as 225,000 barrels per day by 2020 under the assumptions of 5 percent per capita growth, no improvement in transformation losses, and no substitutions of natural gas for crude in electricity generation. If, as seems likely, Cuba switches electricity generation to natural gas, petroleum demand will be some 100,000 barrels per day less than their initial forecast.

Critical Factors in the Development of a Sustainable Cuban Energy Sector

There is certainly no guarantee that the development of Cuban oil and gas reserves will include the participation of U.S.-based oil companies, banks, or government agencies, but a careful review of the possible role for these entities is warranted: what opportunities exist for them and how they might assist the Cubans in achieving energy self-sufficiency while enhancing U.S. geostrategic and energy-security concerns. This is the place to discuss what elements are needed to make that possibility a reality.

Certain conditions are prerequisites for the design, development, and implementation of a policy of promoting energy development cooperation between the United States and Cuba that benefits both countries.

Prioritization of Strategic Goals and Objectives

The most important factor is the clear identification and prioritization of strategic goals and objectives, tasks, and probable outcomes for the development of Cuban energy resources (oil, gas, biomass, solar, wind, and so forth), and the development of energy production capabilities, with special attention paid to the long-term sustainability and impact of these development schemes.[5]

The promotion of energy cooperation implies that there will be relatively open energy markets in Cuba for foreign direct investment that are amenable and accessible to global energy market practices. The prevailing joint-venture (*empresa mixta*) model of investment in Cuba's energy sector has been successful and has significant applicability, should international oil companies based in the United States be offered the opportunity to enter the Cuban market.

U.S. Involvement in the Cuban Energy Sector

The ability of U.S.-based actors to conduct business in Cuba is another critical factor. The presence of national and international oil companies from Spain, Venezuela, and Brazil, among others, does not necessarily imply that U.S. firms will be relegated to the sidelines. In fact, most if not all of these firms rely heavily on first-generation U.S. technology for their deepwater oil exploration, yet U.S. trade controls forbid the transfer of these technologies to Cuba. Thus, it stands to reason that the relaxation of these U.S. trade regulations—permitting the transfer of these technologies, and sales of oil and gas services—are an essential precondition for the creation and development of the Cuban energy sector.

Integration of Cuban Market into the Region

To sustain the development of the energy sector it must be integrated into the market of the entire region. At present the market is segmented by NOCs and IOCs with different development schemes and priorities. Currently Cuba enjoys preferential trade arrangements for oil exports with Venezuela, but there is little if any guarantee that a possible successor regime in Venezuela will be willing to honor the present arrangement. Other factors—including oil price fluctuations and the integrity of Venezuelan oil production rates—might undermine the prevailing order and return Cuba to a situation similar to that in the early 1990s. A loss or dramatic reduction of the supply of Venezuelan oil, though not nearly as catastrophic as the loss of Russian oil in 1992, would create an economic crisis and bring about the termination of significant oil infrastructure projects currently under way (oil pipeline and storage facilities, refineries, petrochemical processing, and power generation plants).

The Cuba-Venezuela Relationship

Cuba's dependency on the Venezuelan relationship is necessary at the moment, but it is not sufficient to meet the requirements of the five "S" characteristics for a secure energy sector in the future, as delineated in chapter 1—especially energy supply, surety, and sustainability. Future optimal development of Cuba's energy sector requires the reduction or elimination of Cuba's dependency on Venezuela for its energy supply, investment capital, and the transfer of technology. The possibility of successful offshore oil and gas production would do much to eliminate this source of uncertainty and alleviate Cuba's overreliance on Venezuela.

Form of Government and the Energy Sector

The aspects of the Cuban energy future explored in the book have not indicated that the successful development of Cuba's energy sector is dependent on a specific form of government or regime type. Governments of all types have successfully exploited their nation's energy resources. It should be noted that many of these governments—for example, Nigeria, Venezuela, and Mexico—have been less successful at managing capital drawn from oil profits because of corruption, cronyism, a lack of transparency, and overly ambitious social and economic development programs. Critical to building a sustainable energy sector is the ability of state energy policy to manage the sector's response to global fluctuations in the price of oil, minimize the impact of regional weather disasters, and enhance the integrity of the sector's management structure to respond to these and other events, such as an oil spill.

International oil companies adapt their practices to market conditions. It stands to reason that if Cuba develops its oil and gas industry via a concession-based model, then IOCs will respond appropriately. If auxiliary (downstream) service firms are required to submit to the *empresa mixta* (joint-venture) model of investment, they will adjust their market strategy to meet that demand.

Since 1992, there has been a significant alteration of energy development priorities in Cuba. The Cuban state has gone from a highly energy-dependent client of the former Soviet Union, enjoying preferential trade arrangements for oil imports and the transfer of advanced nuclear technology for energy development, to an energy-starved state in economic crisis. Cuba is a state that is still relatively dependent on one source, Venezuela, for its energy imports, but has significantly boosted its domestic oil production capacity, and stands at the threshold of a significant transformation if it accesses its substantial offshore reserves. In short, the lesson that can be drawn by observers from the post–cold war history of Cuba is one of incredible adaptability in the face of daunting obstacles, including an almost complete loss of energy supply, the ongoing deterioration of energy infrastructure (transmission and delivery systems), and the demise of energy generation facilities. That the Cuban energy sector has remained relatively stable and responsive to these significant changes is remarkable and bodes well for the new challenges that the country will face in the near and long term.

Cuba, the United States, and the Five "S" Factors

In chapter 1, I spotlighted five "S" characteristics of energy security and the related imperatives of strategic energy policy relevant to both the Cuban case as well as that of the United States. The successful development of Cuban energy resources will enhance the energy security of the United States and its broader geostrategic imperatives in the Caribbean region. Cuba can do this by potentially serving as an *entrepôt* for U.S. downstream activities (refining, marketing, storage, and transshipment). Cuba has already embarked on an aggressive program of investment and development of its refining capacity, which could potentially support American energy needs by serving as a hedge against supply disruptions of refined petroleum products or facilitating the redirection of oil shipments as needed owing to any number of circumstances.

These capabilities could even meet some short-term U.S. market demands. A case in point is the loss of U.S. refining capacity due to damage from Hurricane Rita in the Houston area in 2005. The heavy concentration of U.S. oil infrastructure in the Gulf of Mexico region makes hurricanes of Rita's intensity very problematic. Very little spare crude oil refining capacity exists in the United States. The Gulf of Mexico produces some 2 million barrels per day total, as well as having some 30 percent of the total refining capacity of the United States.[6] Rita's offshore path traveled through an area dense with pipelines and oil platforms, and skirted an onshore area with large refineries.[7] The damage to U.S. refining capacity would have been devastating if Rita had directly hit the Houston region. In the future, the presence of an expanded refining capacity in Cuba might preempt a supply disruption. Alternatively, because Cuba also lies in the path of these tropical storms, having access to American markets provides the Cuban regime with another alternative to respond to and minimize the impact of such events.

A Third Way for the Energy Sector

There are three possible models for the Cuban energy sector: not only the "business as usual" and total marketization models, but also a hybrid "mixed market" scenario. The premise underlying this scenario is that the Cuban state will be the central economic actor in the energy sector. State firms will include Cupet, Unión Eléctrica, and a growing number of state enterprises. The Cubans will seek aggressively to expand their investment and development capabilities in the energy sector in a variation on the Chinese concession-driven economic development model. For example, potential investors are invited to bid on providing the Cuban energy market with the lowest price of electricity, regardless of the mode of generation.

This model can also be applied to oil and gas production, the construction of energy facilities, and associated projects. The model assumes that competition will be vigorous to tap the North Cuba Foreland Basin offshore reserves, for both downstream and upstream potential. Most domestic and international experts believe that the price of electricity may be skewed too low and that the possibility of a low profit rate and long payback terms might dissuade international finance investors from entering the market, or they might quickly exit the market, as has occurred in the China market.[8] Nonetheless, the Cuban state as a joint-venture partner has consistently tried to strike a balance between these concerns from investors to conflate the terms of payback so that the return on investment is more attractive for potential partners.

Cuba will continue to require state majority ownership in all of its joint-venture enterprises (*empresas mixtas*) but there may be the possibility of fuller marketization. Now is the time to begin considering how to design and implement an integrated policy to enhance regional energy security. Oil is traded as a fungible global commodity with little weight given to regime type or ideology. This is especially relevant in the Caribbean context—hurricanes do not recognize national borders nor corporate logos. Thus, waiting for optimal conditions is perhaps inadvisable in the face of the consequences of failing to act.

Conclusion and Recommendations

Undoubtedly, after fifty years of enmity, there is a significant lack of trust and confidence between the United States and Cuba. This is plain from the almost quaint maintenance of a sanctions regime that seeks to isolate Cuba economically and politically but hardly reflects the dramatic changes that have occurred on the island since 1991, not to mention since 2008, when Fidel Castro officially stepped aside as Cuba's president. Now, the opportunity to advance relations in the energy arena appears to be ripe. Since 2004, representatives from American companies, trade organizations, universities, and think tanks have had the opportunity to meet with Cuban energy officials. The scope and objectives of Cuban energy development schemes have been disseminated, dissected, and discussed across a number of settings where the interested parties are now familiar with and well versed in the agendas and opportunities that exist in this arena. In public discussions, Cuban energy authorities have made it clear that their preferred energy development scenario includes working closely with the U.S. oil and gas industry and using state-of-the-art U.S. oil technologies. The assessment from U.S. energy experts on the technical acumen and capability of Cuban energy officials has been overwhelmingly positive.[9] Should the U.S. government and the Obama administration see fit to shift its policy so as to allow broader participation of American academics and practitioners in the energy field to attend conferences and meet with Cuban energy officials, it may pave the way to establishing much-needed familiarity and confidence across these communities.

The United States and Cuba will have a unique opportunity to employ a highly educated and competent cadre of Cuban engineers and technicians to work in critical areas of the energy sector. This will deploy an underused segment of the Cuban workforce, and allow U.S. oil, construction, and engineering firms to subcontract work to an emerging class of Cuban firms specializing in these areas. The Cubans have accumulated experience and training from past energy cooperation projects and exchanges in Cuba, Mexico, Venezuela, and other countries in the region. Anecdotal evidence suggests that these contacts and exchanges have been wildly successful because of the Cubans' high level of competence and strong work ethic. The Cubans have gained invaluable knowledge and experience through the operation and construction of energy facilities in collaboration with their joint-venture partners on the island.

The United States possesses few options when it comes to balancing the various risks to U.S. energy security and satisfying energy demand, because U.S. energy independence is not attainable, the policy tools available to deal with energy supply disruptions are increasingly inadequate, and the United States needs to articulate a new vision of how best to manage international energy interdependence. In particular, even if the United States were to choose to exploit all of its domestic energy resources, it would remain dependent on oil imports to meet its existing and future demand. The critical need to improve the integrity of the U.S. energy supply requires a much broader, more flexible view on the quest for resources—a view that does not shun a source from a potential strategic partner for purely political reasons. U.S. decisionmakers must look dispassionately at potential energy partners in terms of the role they might play in meeting political, economic, and geostrategic objectives of U.S. energy security. The Obama administration has signaled that it wants to reinvigorate inter-American cooperation and integration; a movement toward energy cooperation and development with Cuba is consistent with, and may be central to, that objective.

The energy-security environment for the United States is at a critical juncture. The productive capacity of two of the United States' largest oil suppliers, Mexico and Venezuela, has declined, and the supporting energy infrastructure in both countries is in need of significant revitalization. The vagaries of the politics in the region, the variability of weather patterns, and the overall dismal state of the global economy create a setting of instability and uncertainty that requires close attention to the national security interests of the United States vis-à-vis energy. Cuba's energy infrastructure, too, is in need of significant repair and modernization (its many energy projects notwithstanding); the price tag is estimated to be in the billions of dollars. Delaying work on many of these projects increases costs, because deterioration of the infrastructure continues and eventually pushes up the cost of renovation and replacement. It also stands to reason that the lion's share of the financial burden of upgrading Cuba's energy infrastructure will fall to the United States, directly and indirectly. Changes in U.S. policy to allow investment and assistance in Cuba's energy sector are a precondition for international entities to make significant investments, yet this change implies a large American footprint. Trade and investment in the energy sector in Cuba have been severely constrained by the conditions of the embargo placed on the Cuban regime. These constraints also affect foreign firms seeking to do business in Cuba because of the threat of penalties if any of these firms use technology containing more than 10 percent of proscribed U.S. technologies needed for oil and gas exploration and production. American private investment and U.S. government assistance will constitute a large portion of the needed investment capital to undertake this colossal effort. The longer that work is delayed, the higher the cost to all the investors, which will then potentially cut into the returns from such undertakings.

U.S. cooperation with Cuba in energy just may create an opportunity for the United States to improve its relations with Venezuela, if it can demonstrate that it can serve as a partner (or at a minimum, a supporter) of the Petrocaribe energy consortium. The United States could provide much-needed additional investment capital in the development of upstream, downstream, and logistical resources in Cuba that simultaneously addresses Petrocaribe objectives, diversifies regional refining capacity, and adds storage and transit capabilities while enhancing regional cooperation and integration modalities. This does not mean that the United States has to dismantle the nearly fifty-year-old embargo against Cuba, but the United States will have to make special provisions that create commercial and trade openings for energy development that serve its broad geostrategic and national security goals, as it has in the case of food and medicine sales to Cuba.

This discussion is intended to help distill understanding of U.S. strategic energy policy under a set of shifting political and economic environmental conditions in Cuba and its implications for U.S. foreign policy for the near and long term. Because the policies can be considered works-in-progress, an understanding of possible outcomes is important to those crafting future policy and making changes in the policymaking milieu.

Preconditions for Energy-Sector Cooperation

Four developments have been central to creating the possibility of bilateral U.S.-Cuban commercial cooperation to develop energy resources on the island.

First, Cuba's energy development policy since the economic debacle of the early 1990s has been relatively successful. In that period, it has tapped and expanded existing onshore oil reserves via joint-venture projects undertaken with Canadian firms led by Sherritt. This includes the development of cogeneration facilities in Varadero and Boca de Jaruco.

Second, Cuba landed a second highly favorable energy partner in Venezuela. Venezuela has provided Cuba with over 50 percent of its energy supply through preferential trade arrangements whereby Cuba receives refined fuels in exchange for the deployment of Cuban physicians in Venezuela.

Third, as an extension of its relationship with Venezuela, Cuba has seen a large-scale investment in its energy infrastructure (both actual and planned) valued at billions of dollars. As Cuba expands its refining capacity and replaces refined products with Venezuelan crude oil, it decreases the cost of importing refined fuels, estimated at almost $2 billion in 2008. This growing refining capacity means that Cuba oil exports (primarily refined fuels exported to Petrocaribe partners) now account for 22 percent of Cuba's export earnings, second only to nickel export earnings, at 39 percent.[10] The chance of increasing the percentage is certainly within the realm of possibility as Cuba continues to expand its infrastructure and production capabilities, especially when the offshore reserves are brought on line.

Fourth, in the five years since the announcement of the discovery of offshore oil reserves in December 2004, Cuba has attracted nearly a dozen international and national oil companies interested in leasing offshore tracts in its exclusive economic zone (EEZ) because of the potential of oil and gas resources, which will dramatically alter Cuba's energy balance; Cuba will become a net energy exporter.

Concrete Measures to Promote U.S.-Cuba Energy-Sector Cooperation

There is a significant commercial opening for the United States in Cuba, should it choose to pursue it. Cuban energy development will proceed with or without U.S. involvement, but U.S. involvement has the potential of speeding up the pace of development and could create an opening for a broader discussion of important geostrategic concerns for the both countries. To that end we make the following policy recommendations aimed at facilitating the promotion of strategic commercial relations between the United States and Cuba to develop energy resources.

Recommendation 1. Initiate Confidence-Building Measures and Promote Engagement between the United States and Cuba

Regardless of recent history, the installation of new administrations in both Cuba and the United States creates an opportunity for new modes of engagement that while initially are symbolic can over time pay significant dividends. Strategic energy and infrastructure cooperation may be a relevant consideration for this type of engagement. Prior to any formal cooperation, the United States and Cuba should engage in familiarization and confidence-building meetings and workshops between relevant parties from both countries. Initially these encounters will serve highly symbolic purposes, but over time can become vehicles for information sharing, exchange and assessment regarding mutual goals, objectives, strategies, and policies regarding energy and infrastructure cooperation and development. There are already a number of annual meetings, conferences, and workshops in Cuba that Americans could attend to gain much-needed information regarding Cuba's energy sector. Conversely, the United States government should facilitate the attendance and participation of Cubans in American conferences, exchanges, and residencies.

Recommendation 2. Create Opportunities to Leverage Cuban Human Capital Resources

As a follow-up to recommendation 1, the United States should purposefully create opportunities to leverage Cuba's considerable human capital resources. This would include creating opportunities for internships with U.S. government agencies related to energy resources development, management, and regulation; short-term employment opportunities with U.S. engineering, infrastructure development, and construction firms; and energy-related academic and scientific exchanges. This would be highly beneficial to a whole host of energy-related activities that would be created as a function of foreign direct investment, development assistance, and joint-venture enterprises between the United States and Cuba. Many individuals in Cuba's highly trained cadres of engineers and technicians are underemployed; engaging Cuba in the areas of energy and infrastructure development may provide opportunities to put these people to work and potentially to leverage the considerable skills and abilities of these resources for cooperative projects across the region.

Recommendation 3. Facilitate the Transfer of Critical Energy Technology

As mentioned often throughout this book, the potential of Cuba's offshore oil reserves may only be accessible when Cuba and its partners are able to employ first-generation American deepwater exploration technologies. This is especially critical as many of the firms currently conducting exploration in Cuba's offshore tracts—Repsol S.A. (Spain), Norsk Hydro (Norway), and Petrobras (Brazil)—are also operating platforms under contract to U.S. firms in the Gulf of Mexico. Simultaneously, however, these firms are prohibited from employing first-generation technology from these platforms for their operations in Cuba. The United States should take the steps necessary to ensure that U.S. firms and their subsidiaries are able to employ the technology best suited for the extraction of oil and gas from these deepwater resources. This increases the viability of the operations, avoids costly delays in the operation of these platforms, and enhances the environmental integrity of these operations.[11] At present, U.S. export controls limit everyone's access to this technology. Under more favorable conditions, the United States should begin to roll back export control restrictions in this area as part of energy resource development and production-sharing scenarios.

Recommendation 4. Enhance Cuba's Project Management Capabilities

One of the most critical findings from the analysis of Cuba's effort to develop a nuclear energy capability was the absence of project management capacity during the design, implementation, and construction of the nuclear reactor site at Juragua.[12] Subsequent discussions with senior Cuban government officials revealed that the development of this capacity is a high priority for Cuba as it considers challenges in terms of infrastructure and large construction projects into the future. Cubans have expressed the desire to work side by side with American partners in this critical area. Facility development is an area in which U.S. firms can and should play a vital role as models and partners for Cuba.

Recommendation 5. Promote Energy-Sector Trade and Cooperation

There are numerous areas in the energy sector in which the United States and Cuba can and should cooperate: exploration, energy production, downstream operations, transportation, and auxiliary services. In addition, the opportunity for U.S. firms to invest directly in the development of Cuban energy resources should be created. Recent history shows that Cuba possesses the potential to be a strong regional trade partner in the area of energy and infrastructure development. It might be premature for U.S. firms to begin opening branch offices along Avenida Quinta in the Miramar district of Havana, but it is rational to consider the benefits that would be drawn from the expansion of trade and cooperation between Cuba and the United States in energy development. There will be obvious limitations on such investment opportunities because of the *empresa mixta* joint-venture model that the Cuban government employs, but as previously stated, international oil companies are not averse to adjusting their investment models to specific market conditions, and in the case of Cuba it would be no different. In fact, there has been no lack of interest on the part of American international oil firms in developing a Cuban market. The prevailing Cuban model of joint-venture investment and cooperation has proved to be attractive internationally, and there are numerous opportunities in this area for American firms as soon as there are significant changes in the Cuban embargo so that this type of engagement can occur.

These recommendations establish the basis for developments that speak directly to the enhancement of two broader geostrategic considerations for U.S. energy security: the diversification of regional energy resources and the establishment of a Cuban energy *entrepôt.* The development of partnerships in refining, storage, and engineering services will allow the regional partners to diversify their respective portfolios, in addition to dispersing resources across the region to take advantage of location, and perhaps mitigate potential market disruptions owing to weather and other natural disasters.

A further long-term prospect for Cuba may be the development of energy-related resources that are positioned strategically to serve the region in terms of oil refining and storage, oil and gas production (exploration and infrastructure), and auxiliary services. These developments would be a boon to Cuban, American, and regional economic development interests and are especially relevant in the context of growing concerns over the energy infrastructure in the region, and in particular the oil and gas industries of Mexico and Venezuela.

Oil exploration is an inherently risky enterprise; there are always trade-offs between negatives and positives relating to energy security, environmental integrity, and geostrategic considerations. The consensus arising from the studies and the analyses in this book is that the creation of mutually beneficial trade and investment opportunities between the United States and Cuba is long overdue. Throughout most of the twentieth century, Cuban infrastructure and economic development were direct beneficiaries of commercial relations with the United States. This relationship was instrumental in providing Cuba with access to advanced technologies and the signs of modernity that were unparalleled in Latin America and far beyond.

Once again, the United States is presented with an opportunity that might serve as the basis of a new relationship between the United States and Cuba. It holds out the possibility of enhancing the stability and development of a region that is wrestling with questions of how and when it too might benefit from engagement with a global economic development model. The question is whether the United States chooses to be at the center, or to leave Cuba to seek some alternate path toward its goals.

Ironically, Cuban officials have invited American oil companies to participate in developing their offshore oil and natural gas reserves. American oil, oil equipment, and service companies possess the capital, technology, and operational know-how to explore, produce, and refine these resources in a safe and responsible manner. Yet they remain on the sidelines because of our almost five-decades-old unilateral political and economic embargo. The United States can end this impasse by licensing American oil companies to participate in the development of Cuba's energy resources. By seizing the initiative on Cuba policy, the United States will be strategically positioned to play an important role in the future of the island, thereby giving Cubans a better chance for a stable, prosperous, and democratic future. The creation of stable and transparent commercial relations in the energy sector will bolster state capacity in Cuba while enhancing U.S. geostrategic interests, and can help Cuba's future leaders avoid illicit business practices, minimize the influence of narcotrafficking enterprises, and stanch the outflow of illegal immigrants to the United States.

If U.S. companies are allowed to contribute to the development of Cuba's hydrocarbon reserves, as well as the development of alternative and renewable energy (solar, wind, and biofuels), it will give the United States the opportunity to engage Cuba's future leaders to carry out long-overdue economic reforms and development that will perhaps pave the way to a more open and representative society while helping to promote Cuba as a stable partner and leader in the region and beyond.

Under no circumstances is this meant to suggest that the United States should come to dominate energy development policy in Cuba. The United States certainly has a role to play, but unlike its past relationship with Cuba, its interaction and cooperation will be predicated on its ability to accept, at a minimum, that Cuba will be the dominant partner in potential commercial ventures, and an equal partner in future diplomatic and interstate relations. Without a doubt Cuban government actors are wary of the possibility of being dominated by the "colossus of the North," but as Cuba's energy policy-makers face the daunting reality of their nation's energy future, it is abundantly clear that they possess the willingness and the capacity to assiduously pursue sound policy objectives and initiatives that begin to address the island's immediate and long-term challenges. In the end, this course of action will have direct and tangible benefits for the people of Cuba, it neighbors, and beyond.

Notes

1. Mahmoud Amin El-Gamal and Amy Myers Jaffe, "Energy, Financial Contagion, and the Dollar," working paper, Working Paper Series: The Global Energy Market: Comprehensive Strategies to Meet Geopolitical and Financial Risks (Houston: Rice University, James A. Baker III Institute for Public Policy, Rice University, 2008), p. 26.

2. According to daily data at the website OilPrice.net, as of February 2010, the price of a barrel of oil had ranged between $78.25 and $80.22 for the preceding year (www.oil-price.net/index.php?lang=en). This is the actual price and not the forecast price.

3. The potential role of bagasse depends on the future direction taken by Cuba's sugar industry.

4. Cuba has begun to replace its electricity delivery infrastructure in the greater Havana metropolitan area. This has helped to reduce the distribution losses from 18 percent in 2005 to 14 percent in 2009. Unión Eléctrica, the Cuban state electricity utility, has set 11 percent distribution losses as its goal for 2012 although the 29 percent total losses of electricity in transmission and delivery is evidence of the critical need to boost efficiency. Author interview with Juan Fleitas Melo, assistant to the minister of basic industry, Havana, Cuba, November 18, 2009.

5. Cuba's Ministry of Science, Technology and the Environment (CITMA), has undertaken extensive international research and outreach efforts to address these questions and actively hosts international conferences to explore the issues related to energy, technology transfer, and sustainability.

6. According to a former oil industry executive, Jorge Piñón, in September 2005, when over half of U.S. Gulf oil production was still shut down in the wake of Hurricane Katrina, some economists outlined a worst-case scenario in which gasoline prices reached $5 per gallon. In fact, the oil industry emerged essentially unscathed from the storm, and post-storm price rises were minor. Nonetheless, the economic

impact of the storm was significant, estimated by some as a 1 percent reduction in the GDP growth rate in the second half of 2005, from job losses, reduced production at refineries, and disruptions to the energy supply. Author interview with Jorge Piñón, New York, February 21, 2008.

7. See A. M. Cruz and E. Krausmann, "Damage to Offshore Oil and Gas Facilities Following Hurricanes Katrina and Rita: An Overview," *Journal of Loss Prevention in the Process Industries* 21, no. 6 (November 2008): 620–26.

8. Charlie Dou, "Attractive Prospective and Uncertainties of China Wind Power Market," *GLG News,* March 19, 2007 (www.glgroup.com/News/Attractive-prospective-and-uncertainties-of-China-wind-power-market-9509.html). The author is the CEO of Beijing Bergey Windpower Company.

9. This conclusion is based on a series of informal discussions with U.S. attendees at conferences and meetings since 2004—including events in Mexico City, Havana, New York, and Miami—where materials and presentations on the Cuban energy sector were disseminated.

10. Mark Frank, "Oil Now Second-Leading Cuban Export—Gov't Report," *Reuters,* June 10, 2009.

11. In the period since 2000 no fewer than three hundred platforms have been damaged and taken offline because of hurricanes and other weather-related causes, yet no catastrophic events have occurred as a result of using state-of-the-art U.S. deepwater extraction technologies at these platforms. Oil-related incidents that had occurred had been mostly in the area of marine transportation of oil and other petroleum products in and near ports. In the wake of the April 2010 BP-Deepwater Horizon disaster in the northern Gulf of Mexico, it remains to be seen what type of new measures for deepwater oil exploration will be instituted and whether the United States will seek to extend these measures to regional and international oil producers.

12. Jonathan Benjamin-Alvarado, *Power to the People: Energy and the Cuban Nuclear Program* (New York: Routledge, 2000).

Contributors

Juan A. B. Belt is director of Chemonics International, an international development consulting company, and is former director of the Office of Infrastructure and Engineering at the U.S. Agency for International Development and has worked for the Inter-American Development Bank and the World Bank.

Jonathan Benjamin-Alvarado is a professor of political science at the University of Nebraska–Omaha and the assistant director of the Office of Latino/Latin American Studies. He is also senior research associate of the Center for International Trade and Security at the University of Georgia.

Amy Myers Jaffe is the Wallace S. Wilson Fellow in Energy Studies, James A Baker III Institute for Public Policy, Rice University, and served as a member of the reconstruction and economy working group of the Baker/Hamilton Iraq Study Group and as project director for the Baker Institute/Council on Foreign Relations task force on Strategic Energy Policy.

Jorge R. Piñón is an international energy consultant and visiting research fellow with Florida International University's Cuban Research Institute. He was president of Amoco Oil Latin America based in Mexico City and general manager of BP Europe's Western Mediterranean Supply and Logistics organization based in Madrid.

Ronald Soligo is a Rice Scholar at the James A. Baker III Institute for Public Policy, Rice University. He is currently working on the potential for ethanol exports from Caribbean countries.

Index

Agriculture sector (Cuba), 85–86, 95, 98–99, 104–05, 115. *See also* Sugar industry (Cuba)
Alaska Natural Wildlife Reserve (ANWR), 2
ALBA (Alternativa Bolivariana para las Americas). *See* Bolivarian Alternative for the Americas
Alonso-Pippo, Walfrido, 66, 67
Alternativa Bolivariana para las Americas (ALBA). *See* Bolivarian Alternative for the Americas
Angola, 6, 32
ANWR. *See* Alaska Natural Wildlife Reserve
"Assessment of Undiscovered Oil and Gas Resources of the North Cuba Basin 2004" (report; U.S. Geological Survey), 31
Automobile industry, 90, 103

Bader, Jeffrey, 12–13
Bay of Cardenas, 83
Belt, Juan A. B., 14, 48–79, 113
Benjamin-Alvarado, Jonathan, 1–20, 21–47, 84, 110–29
Biomass and bagasse, 62, 63, 64, 66–68, 103–04, 113
Blanco, Ramon, 32
Boca de Jaruco (Cuba), 123
Bolivarian Alternative for the Americas (*Alternativa Bolivariana para las Americas*; ALBA), 9, 12
Brazil: food and environmental issues of, 115; as Cuban regional partner, 4, 6, 11; energy and hydrocarbon sectors in, 94,106, 111, 116, 125; soybeans and, 100; sugarcane and ethanol industry in, 80, 94, 97, 100, 103, 115; visit of R. Castro to, 6. *See also* Petrobras; Proálcool
Brookings Institution, 4
Brufau, Antonio, 32
Buckskin Project (Gulf of Mexico), 33
Bundling and unbundling, 71, 75, 113

Canada, 26, 28. *See also* Sherritt International

Castro, Fidel, 7, 50, 94, 97, 120
Castro, Raúl, 6, 7, 9, 15
Central America, 103. *See also individual countries*
Cereijo, Manuel, 55
Chávez, Hugo, 12
Chile, 48–49, 54, 58, 59t
China: cooperation agreement with Cupet, 83; as crude oil importer, 12; economic transition of, 87; growing influence of, 9, 12; investment in China, 119–20; investment in Cuba, 82; Latin American energy markets and, 18, 111; oil companies of, 32; Russian loan of, 36–37; U.S. energy security and, 13
China Petroleum and Chemical Corporation (Sinopec), 27, 83
Chinese National Offshore Oil Corporation (CNOOC), 32
Cienfuegos (port city; Cuba), 29, 36, 78n14, 81, 112, 113
CMEA. *See* Council for Mutual Economic Assistance
CNOOC. *See* Chinese National Offshore Oil Corporation
Coal, 37
Cold war and post–cold war period, 1, 5, 8, 36, 37, 118
Colombia, 31
Commercial sector (Cuba), 84, 88
Communism and Communist bloc, 48
Competition, 71
Contractors, 24, 121
Costa Rica: electric power data of, 48–49, 54, 58–59; energy sector of, 87–88, 89, 92
Council for Mutual Economic Assistance (CMEA), 95, 96
Cuba: Cold war exports of, 36; cooperation with the U.S. and, 2–4, 16–17, 123–27; Cuban-U.S. energy policies, 7, 114, 123–26; electric power challenges in, 22; human capital resources of, 3, 16, 124–25; leadership role of, 3; maritime boundary between Cuba and the U.S., 31; national and international policy challenges of, 5–6, 11, 22; population of, 92; project management capabilities of, 125; recommendations for, 15–17; regional partners of, 4; stability of, 4; tourism in, 28, 29, 82, 90, 105–06, 112; work ethic and competence in, 121. *See also* Castro, Fidel; Castro, Raúl; Economic issues (Cuba); Electric power sector (Cuba); Energy sector (Cuba); Exclusive Economic Zone; Hydrocarbon sector (Cuba); Unión Cubapetróleo S.A.; United States
Cuba—individual countries: Angola, 32; Brazil, 33, 34, 35, 36, 37, 83; Canada, 26, 27, 28, 32, 51; China, 27, 32, 33, 36–37, 82, 83; Germany, 53; India, 32–33, 35, 83; Malaysia, 35; Norway, 32, 33, 34, 35, 37, 83; Russia, 18, 27, 36–37, 111; Soviet Union, 6, 22, 36, 37, 49, 50, 81, 84, 88; Spain, 32, 33, 34–35, 37, 53, 83; Tobago, 29; Trinidad, 29; Vietnam, 27, 35–36; West Germany, 53. *See also* United States; Venezuela
Cuban Liberty and Democratic Solidarity (Libertad; Helms-Burton) Act of *1996*, 33, 34, 76
Cuban Revolution (*1959*), 93
Cupet. *See* Unión Cubapetróleo S.A.

Dallas Morning News, 1
Deepwater Horizon oil spill (*2010*), 33, 34, 129
Defense, Department of (U.S.), 10
Direct foreign investment (DFI). *See* Foreign direct investment

Dominican Republic: electric power data of, 48–49, 54; energy sector of, 87–88, 89, 92; imports of LNG of, 29; productivity of power sector of, 58–59
Drilling: bidirectional drilling technology, 27; deepwater drilling, 33, 38, 83; directional drilling technology, 26, 27; environmental issues of, 33–34; exploratory drilling, 31, 32–33; slant drilling technology, 83

Earnings before interest, taxes, depreciation, and amortization (EBITDA). *See* Economic issues—Cuba
Eastern Europe, 87, 96
EBITDA (Earnings before interest, taxes, depreciation, and amortization). *See* Economic issues—Cuba
Economic issues: bundling and unbundling, 71, 75, 113; commodity markets, 94; competition, 69, 71; ethanol prices, 101; foreign exchange, 95, 97; global economy, 121; import substitution strategies, 95; monopoly, 69–70, 71; oil trading, 120; recovery of undiscovered recoverable reserves, 92–93; shifts in economic development model, 87; single purchasing agency, 70; sugar prices, 101
Economic issues (Cuba): alternative scenarios, 13–14; commodity markets, 82, 94–95; dual monetary system, 55; EBITDA, 26, 55–56; economic crisis of *2008*, 26–27; economic development model, 87, 114; energy profile, 84, 105–06; foreign exchange and exchange rates, 60, 97, 105–06; gross domestic product (GDP), 15, 51, 81; market strategies of oil companies, 118; oil exports, imports, and trade, 36, 40, 49, 82, 84, 95, 105–06, 115, 123; per capita income, 81, 87, 90, 107n22, 107n23; *periodo especial,* 50–51, 114; price of oil and, 56–58; privatization, 49, 50; Soviet Union and Russia and, 37, 81, 84, 114, 117; sugar industry, 94–97; supply and demand for petroleum, 41, 44, 90; tourism, 82; Venezuela and, 45t, 50–51, 81, 117; wages and costs of labor, 55; worldwide recession and, 82. *See also* Energy sector (Cuba); Foreign direct investment; Hydrocarbon sector (Cuba); Unión Eléctrica; United States—sanctions and embargo against Cuba
Ecuador, 31
EEZ. *See* Exclusive Economic Zone
Electric power sector: costs of, 53; distributed generation, 53; ethanol and, 103; models of, 69–71
Electric power sector (Cuba): blackouts and, 51, 53, 70, 86, 88; control and management of, 48, 49; costs of, 53–54, 55–56, 59–61, 62–63, 64, 68, 113; demand for, 62, 64; electric appliances in Cuba, 88; environmental issues of, 65; future development and investment prospects and scenarios for, 60–65, 68–76; gensets, 51, 53, 61; *grupos electrógenos,* 53–54; incentives for, 53; increasing efficiency and sustainability potential, 68–69; main periods and trends, 50–54; modernization of, 71–76, 90; nationalization of, 50; power consumption, 48, 49t, 50, 51, 52, 54; price of, 119; production of power, 49, 50, 51, 52, 53, 61, 62–65, 113, 115–16; regulation of, 50, 73; replacement of existing plants, 13,

63, 64–65, 74; *revolución energética,* 51, 53; Soviet Union and, 50, 51; structure of power sector, 70f, 71–72; subsidies for, 51, 63, 64; tariffs for, 61, 63, 68, 69, 72, 74, 113; value of energy consumed, 49, 55; Venezuela and, 51, 55–56, 61, 64. *See also* Cuba; Energy sector (Cuba); Ethanol industry (Cuba); Hydrocarbon sector (Cuba); Renewable energy sector (Cuba); Unión Eléctrica

El-Gamal, Mahmoud Amin, 10, 110

El Salvador, 70

Employment and labor, 97, 99, 104, 121

Empresa mixta. See Joint ventures

Energas (power company; Cuba), 28, 76, 83. *See also* Sherritt International

"Energy, Financial Contagion and the Dollar" (paper; El-Gamal and Jaffe), 110

Energy Independence and Security Act of *2007,* 97

Energy Information Administration (U.S.), 25, 28, 63

Energy sector: breakdown of energy use, 87–88; costs of, 53, 90; distributed generation, 53; energy balances, 107n18; energy demand, 90; energy efficiency, 86; strategic energy policies, 110–11

Energy sector (Cuba): analytical framework and scenarios for Cuban energy, 13–17; competition, 12; concrete measures to promote U.S.-Cuba cooperation, 123–26; critical factors in the development of a sustainable sector, 116–19; distributive energy markets, 3; incentives for, 28, 67, 104; integration of Cuban market into the region, 117; liquid fuels and, 49, 50t, 51, 57, 58, 59, 74, 77n3, 113; natural gas and, 62, 105; nuclear energy capability of, 16; ratio of TFC to TPES, 86, 90; recommendations and reforms for, 113–14, 120–22, 123–28; *revolución energética* and, 51, 53; Soviet Union and, 84, 118; United States and, 116–17, 118–19, 120–28. *See also* Electric power sector (Cuba); Ethanol industry (Cuba); Hydrocarbon sector (Cuba); Sugar industry (Cuba)

Energy sector (Cuba)—specific energy issues: energy balance, 114; energy cooperation with the U.S. and, 2–4, 16–17, 22; energy demand, 12, 90–91, 92, 105, 114, 115–16; energy development, 4–6, 7, 11, 22, 28–29, 38, 82–83, 118–19, 123–24; energy efficiency, 114; energy imports, 13, 81, 123; energy infrastructure, 121–22, 123, 126; energy potential, 13, 80, 105, 114; energy profile, 84–86, 114; energy security, 113, 114, 117, 118–19; energy self-sufficiency, 93, 114, 115; energy theft, 77n11; energy use, 87–88; future energy demand, 86–92; total final consumption (TFC) of, 87–88, 92; total primary energy consumption (TPEC) of, 90, 92; total primary energy supply (TPES), 84, 86, 88–89, 92

Energy sector (U.S.): biofuel use, 97; energy imports, 3, 121; energy independence, 121; energy security, 2, 3, 4, 9–13, 106, 110–11, 114, 118–19, 121, 126; ethanol industry, 98, 102–03, 115; investment in Cuban energy infrastructure, 122; oil imports, 2, 12; oil platforms in the Gulf of Mexico, 125; recommendations, 15–17, 127; refining capacity,

119; sugar imports from Cuba, 93, 95; sustainable energy resources, 5
Energy security. *See* Energy sector (Cuba); Energy sector (U.S.)
Energy transformation losses, 88, 90, 92, 105, 113, 114, 128n4
England, 75
Eni S.p.A. (oil and gas company; Italy), 33
Environmental issues, 2, 29, 33–34, 51, 53, 65, 82
Ethanol industry (Cuba): exports, 115; future of the sugar and ethanol industries, 97–105, 114–15; industrial ethanol output levels, 100–01; issues in achieving Cuba's ethanol potential, 102–03, 115; structure of the ethanol industry, 104–05. *See also* Renewable energy sector (Cuba); Sugar industry (Cuba)
Europe. *See* Eastern Europe
Exclusive Economic Zone (EEZ; Cuba): environmental issues in, 33–34; exploration blocks in, 30, 32, 112; exploratory drilling in, 32–33; location and size of, 29–30; oil and gas reserves in, 13, 21–22, 29–30, 31, 123; oil companies active in, 35–37, 38, 112, 123
Exclusive Economic Zone (EEZ; Florida), 30
Exxon Valdez oil spill (*1989*), 33

FDI. *See* Foreign direct investment
Florida Strait, 21, 29–30, 83. *See also* Gulf of Mexico
Food and food crops, 98–99, 115
Foreign direct investment (FDI): Cuban energy markets and, 116; in critical Cuban infrastructure, 13; Cuban human capital resources and, 124; in the Cuban hydrocarbon sector, 25, 111; in Cuban power sector, 71–72; global financial crisis and, 68, 113; significant projects, 82; U.S.-Cuba energy cooperation and, 2
Foreign private firms, 104

Gazprom (gas company; Russia), 37
Germany, 53. *See also* West Germany
Great Wall Drilling Co. (China), 27
Guatemala, 70, 87, 92
Gulf of Mexico, 29–32, 36, 38, 119, 125. *See also* Florida Strait

Hartley, Peter, 93
Hawaii, 95, 103–04
Helms-Burton Act. *See* Cuban Liberty and Democratic Solidarity (Libertad) Act of *1996*
Housing. *See* Residential sector
Human capital. *See* Cuba
Hunt, Sally, 75
Hurricanes, 53, 119, 128n6, 129n11
Hydrocarbon sector: definition of undiscovered recoverable reserves, 92; oil prices, 90, 108n24, 112, 128n2; recovery of undiscovered recoverable reserves, 92–93; risks of oil exploration, 126; Russia and, 36–37
Hydrocarbon sector (Cuba): combined-cycle gas turbines, 113; direct foreign investment in, 25; energy development agreement with Russia, 111; environmental issues of, 33–34, 51, 53; exported products, 123; future development challenges, 34–38, 62; hydrocarbon production, 2, 25, 26, 27–28, 29t, 39, 82–83; imported products, 42, 49, 62–63, 64, 81, 84, 86, 105; *periodo especial* and, 50; possible models of, 119–20; privatization in, 49; projected annual production,

93, 105, 111; ratio of total consumption to total supply (TFC to TPES), 86, 114; recovery rates of, 25, 27, 28; refining and refineries, 29, 34, 43, 81, 82, 84, 111, 112, 114, 118–19, 123; regasification facility, 29, 112, 113; Soviet Union and, 84, 85; tankers and transport in, 81; technology and, 26, 27–29, 31, 32–33, 92–93. *See also* Energy sector (Cuba); Electric power sector (Cuba); Petroleum; Production-sharing agreements; Venezuela; United States

Hydrocarbon sector (Cuba)—natural gas: characteristics of, 25; costs and prices of, 62–63, 64; Cuban energy profile and, 84, 88t; Cuban reserves and resources, 2, 13, 25, 28–29, 31, 37, 80, 93, 105; financing and construction of new plants, 63; foreign direct investment in, 111; imports of, 29, 42, 62–63, 64, 112, 113; LNG, 29, 62–63, 64, 65, 112; power generation and, 61, 112, 115–16; production sharing agreements and, 24; recovery of, 27; refinery production, 43; total annual supply of, 86; undiscovered reserves of, 92; U.S. trade restrictions and, 35

Hydrocarbon sector (Cuba)—offshore resources: Cuban independence and, 6, 118; Cuba-Canada cooperation, 26, 82–83; Cuba-U.S. cooperation, 4, 127; Cuba-Venezuela cooperation, 117; deepwater technology and, 16, 26, 112, 125; early discoveries, 21; in the EEZ, 29–34, 35–37, 38, 112; environmental issues, 33–34; exploratory drilling, 32–33, 38; geography of oil in the Gulf of Mexico, 30–32; international oil companies and, 112; lease agreements for offshore tracts, 18; oil reserves, 13, 21–22, 38, 82–83, 92–93; production of, 93, 111, 123

Hydrocarbon sector (Cuba)—oil: characteristics and quality of oil, 25–28, 50, 61; crude oil production, 25, 29t, 39, 49, 86, 111–12; crude oil supply, 84, 86; Cuba-Venezuela estimated petroleum debt, 45, 81; demand and supply for oil, 41, 44, 115–16; drilling in the Gulf of Mexico, 27, 32–33; final end-use of oil, 25, 49, 50t; imports, 42, 81; land and marine oil blocks, 22, 23, 83; legal status of oil reserves, 24–25; liquid fuel supply, 49, 50t; refinery production, 43; undiscovered oil reserves, 31, 80, 92–93. *See also* Energy sector (Cuba); Electric power sector (Cuba); Exclusive Economic Zone (Cuba)

Hydrocarbon sector (Cuba)—prices, costs, and value: drilling and exploration costs, 32, 37; economic issues of oil exports, imports, and trade, 36, 40, 49, 82, 84, 95, 105–06, 115, 123; estimated petroleum debt to Venezuela, 36, 37t; financing by Venezuela, 81; market value of petroleum imports, 40; natural gas prices, 62–63, 64; North Coast deepwater project, 33; oil prices, 26, 27, 33, 56–58; price basis for production, 25; production costs, 26; realized crude oil value, 25, 111; value of energy consumed, 49

Hydropower, 65, 113

IAEA. *See* International Atomic Energy Agency

Independent power producers (IPPs), 70, 75–76

India, 32–33, 83
Industrial sector, 85–86, 95
Institute for Public Private Partnerships (IP3), 74, 76
Inter-American Development Bank, 105
International Atomic Energy Agency (IAEA), 78n16
International Energy Agency, 81
International oil companies (IOCs): adjustment of investment models of, 126; adjustment of market strategies of, 118; deepwater EEZ block assignments of, 112; development of energy sector and, 117; joint-venture projects and, 14, 82–83; proven oil reserves and, 2, 111; United States and, 112–13, 116. *See also* Joint ventures; *individual companies*
Inter RAO UES (energy company; Russia), 82
IOCs. *See* International oil companies
IPPs. *See* Independent power producers
IP3. *See* Institute for Public-Private Partnerships

Jaffe, Amy Myers, 10, 14, 80–109, 110, 114–15
Jamaica, 87
Joint ventures (*empresa mixta;* Cuba), 24–25, 116, 118, 120, 121, 123, 126. *See also* International oil companies
Juragua (nuclear reactor site; Cuba), 16, 125

Kennedy School of Government (Harvard University), 73, 76
Kleber, Drexel, 9, 10

Latin America, 73, 92. *See also individual countries*
Lenin, Vladimir Ilyich, 48
Leverett, Flynt, 12–13

Malaysia, 35
Mariel (port; Cuba), 82
MARKAL/TIMES energy systems model, 14, 61–62, 77n13, 113
Medlock, Kenneth, 87, 90, 93
Mesa-Lago, Carmelo, 86, 97
Mexico, 3, 12, 33, 117–18, 121, 126. *See also* Petroleos Mexicanos S.A.
Minbas. *See* Ministry of Basic Industry
Minerals Management Service (U.S.), 30
Ministry of Basic Industry (Minbas; Cuba), 24, 50, 73
Ministry of Communications (Cuba), 50
Monopoly, 69–70, 71
Moreno, Conrado, 94

National Association of Regulatory Utility Commissions, 74
National oil companies (NOCs), 2, 11–12, 35, 111, 117. *See also individual companies*
National Renewable Energy Laboratory (U.S.), 66
Natural gas, 28, 29, 37, 62–63. *See also* Hydrocarbon sector (Cuba)—natural gas
New York Times, 1, 21
Nicaragua, 74, 80
Nickel, 82, 123
Nigeria, 2, 117–18
NOCs. *See* National oil companies
Norsk Hydro (oil company; Norway), 32–33, 35, 125
North Coast Basin, 112
North Coast Reserves (Cuba), 111
North Cuba Basin, 2, 31, 80, 92
North Cuba Fold and Thrust Belt, 25, 31, 37, 46n25
North Cuba Foreland Basin, 31, 46n25, 119
North Sea, 35

Norway, 32, 83,125
Nuclear energy, 16, 118, 125

Obama (Barack) administration, 121
Oil: challenges for development, 34–37; geography of oil in the Gulf of Mexico, 30–32; global demand for, 12–13; global supply of, 13; holdings of proven oil reserves, 2; oil prices, 56–58, 59–60; quality of, 25; risks of exploration, 33–34; undiscovered oil and natural gas, 31. *See also* Hydrocarbon sector (Cuba)—oil
Oil and Natural Gas Corporation (ONGC; India), 32–33, 35, 83
Oil reserves. *See* Hydrocarbon sector (Cuba)—oil; Petroleum
ONGC. *See* Oil and Natural Gas Corporation

PDVSA. *See* Petróleos de Venezuela S.A.
Pebercan (oil company; Canada), 26–27, 83
Pemex. *See* Petróleos Mexicanos S.A.
Periodo especial (special period), 50–51, 98, 114
Petrobras (oil company; Brazil): deep-water expertise and exploration of, 33, 36, 37, 83, 112; ethanol and, 103; government holdings in, 35; platform in the Gulf of Mexico, 34, 125
Petrocaribe oil consortium, 7, 12, 36, 122, 123
Petróleos de Venezuela S.A. (PDVSA), 11, 27, 29, 36, 83, 111, 112
Petróleos Mexicanos S.A. (Pemex; Mexico), 2, 11, 12
Petroleum. *See* Oil
Petronas (oil company; Malaysia), 35
PetroVietnam, 27, 35, 83
Philippines, 95
Piñón, Jorge, 14, 21–47, 67, 93, 111, 112, 114
Pollitt, Brian, 102
Population, 92
Power purchase agreements (PPAs), 53, 70, 77n3
PPAs. *See* Power purchase agreements
Prince William Sound (Alaska), 33
Preferential trade arrangements, 36, 117, 118, 123
Proálcool (national ethanol-production program; Brazil), 103
Production-sharing agreements (PSAs): Canada and, 26; China and, 13; concepts of, 45n4; Cuba and, 24, 26; natural gas and, 24; phases of, 24; privatization and, 49; Spain and, 33; U.S.-Cuba energy cooperation and, 2
Project on U.S. Policy toward a Cuba in Transition (Brookings), 4
PSAs. *See* Production-sharing agreements
Public Service Commission (Cuba), 50
Public Utilities Research Center (Univ. of Florida), 73, 76
Public Utility Commission (PUC; Cuba), 73–75
Puerto Rico, 29, 95

Regulation, 50, 73–74
Renewable energy sector, 94, 103
Renewable energy sector (Cuba): bagasse and biomass, 62, 63, 64, 66–68, 103–04, 113; biofuels, 97, 100; costs of, 65; ethanol, 53, 80, 93–94, 97–105, 114–15; hydropower, 65, 113; solar photovoltaic, 65–66, 113; sources of renewable energy, 65–68; wind power, 53, 63, 64, 66, 113. *See also* Ethanol industry (Cuba); Sugar industry (Cuba)

Renewable Energy Laboratory of the Americas, 5
Renewable fuels standard (RFS), 97
Repsol-YPF (oil and gas company; Spain), 32, 33, 34–35, 37, 83, 112, 125
Residential sector (Cuba), 84–85, 88, 90
Rethinking U.S.-Latin American Relations (report; Partnership for the Americas Commission), 4, 5
RETscreen Clean Energy Project Analysis Software, 66
RFS. *See* Renewable fuels standard
Rice University World Gas Trade Model, 93
Rifkin, Jeremy, 3
Romero, Simon, 1
Rosneft (oil enterprise; Russia), 37
Russia, 6, 36–37, 117. *See also* Soviet Union

Sanchez, Juan, 102
Saudi Arabia, 2
Scandinavia, 75
Sherritt International (oil company; Canada): Cupet and, 26, 27, 28; deepwater block in the EEZ, 32; joint ventures with Cuba, 123; Pebercan and, 26; power production in Cuba and, 51, 53, 70f, 82, 83. *See also* Energas.
Single purchasing agency, 70
Sinopec. *See* China Petroleum and Chemical Corporation
Smith, Warrick, 73
Smoot-Hawley tariffs (*1930;* U.S.), 95
Solar and Wind Energy Resource Assessment project, 66
Solar photovoltaic power, 65–66, 113
Soligo, Ronald, 14, 80–109, 87, 90, 114–15
Sonangol (oil company; Angola), 32
South Korea, 53
Soviet Union: CMEA and, 95, 96; collapse of, 6, 50, 81, 84, 96, 98, 114; Cuban imports of oil and gas from, 36, 37, 50, 84; cut-off of aid to Cuba, 81, 88; economic transition of, 87; electric power sector in, 48; sugar imports from Cuba, 93, 96; Varadero, Cuba oil field and, 22. *See also* Cold war; Russia
Soybeans, 99–100
Spain, 32, 33, 34–35, 37, 53, 75
Special period. *See Periodo especial*
Statoil–Norsk Hydro (oil and gas company; Norway), 33, 34–35, 37, 83, 112
Strategic energy policy, 8, 9, 10, 110–11, 116, 118, 122
Sugar industry (Cuba) : biomass and bagasse and, 62, 63, 64, 66–68, 103–04, 113; biorefineries, 102, 104; collapse of, 86; exports of sugar, 93; future of the sugar and ethanol industries, 97–105; history of sugar cultivation, 94–97; industrial ethanol output levels, 100–01; issues in achieving Cuba's ethanol potential, 102–03, 115; land planted with sugarcane, 98–100, 115; needed investment in, 102–03; prices of sugar, 95, 101; Soviet Union and, 96, 98; sugarcane, 66–67, 100, 102; sugar production, 93–94, 95, 96, 98, 101. *See also* Agriculture sector (Cuba); Ethanol industry (Cuba); Renewable energy sector (Cuba)
Sustainable energy resources, 5. *See also* Renewable energy sector

Taxes, 24
Technology: Cuban EEZ and, 31; deepwater exploration technology, 16, 83,

116–17, 126; directional drilling technology, 26, 27; enhanced oil recovery (EOR) technology, 27, 28; gensets, 51, 53, 61; Petrobras and, 35; ratio of TFC to TPES, 86; recovery of undiscovered recoverable reserves, 92–93; slant drilling technology, 83; United States and, 2, 116–17, 120, 122, 125, 127
Telecommunications (Cuba), 76
Texas, 74
TFC. *See* Total final consumption
Time magazine, 21
Tobago, 112
Total final consumption (TFC; Cuba), 86, 87–88
Total primary energy supply (TPES; Cuba), 84, 86, 88–89
Tourism. *See* Cuba
TPES. *See* Total primary energy supply
Transmission and transmission companies: deterioration in, 118; distributed generator project and, 54; losses incurred during transmission, 86, 114, 128n4; recommendations, 74, 75; structure of Cuba's energy sector and, 70f, 71, 72f; unbundling and, 71; wind transmission, 66
Transneft (oil pipeline monopoly; Russia), 36–37
Transportation sector, 97
Transportation sector (Cuba), 85, 88, 90, 94
Trinidad, 112

Unión Cubapetróleo S.A. (Cupet; Cuba): China and, 27; Canada and, 26–27; Cuba's EEZ and, 112; exploratory work of, 27–28; natural gas and, 28; PDVSA and, 36; PSAs of, 24–25, 26–27, 32, 83; recoverable offshore oil and, 93; regasification facility of, 29, 112; Sherritt International and, 32; United States and, 11, 111, 112. *See also* Energas; Exclusive Economic Zone
Unión Eléctrica (power company; Cuba), 28, 54–60, 70, 74, 75, 82, 113
Union of Soviet Socialist Republics (USSR). *See* Soviet Union
United Nations Environment Program, 66
United States (U.S.): cooperation with Cuba and, 2–4, 16–17, 114, 123–27; Florida EEZ, 30; geostrategic interests of, 2, 9–13; humanitarian aid to Cuba of, 4, 122; Latin American and Caribbean partnerships of, 4; maritime boundary between Cuba and the U.S., 31; Mexico and, 12; national security of, 122; NOCs and, 11–12; recommendations for, 15–17, 127; support for Cuban reforms, 76; U.S. and foreign participation in Cuba, 22–24, 26–28, 32–33, 34, 49, 83; U.S. policies toward Cuba, 1–2, 4–5, 7, 38, 114, 121, 123–26; Venezuela and, 12, 122. *See also* Energy sector (U.S.)
United States—sanctions, tariffs, and embargoes: effects of, 81, 82, 83–84, 99, 105, 112–13, 120, 127; export controls, 16, 125; exports of equipment to Cuba, 50; imports of oil from Cuba, 34; lifting of, 35, 126; recommendations for, 34, 114, 117; sanctions against machinery and technology, 33, 34, 38, 76, 83–84, 112–13, 116–17, 120, 122; sugar quotas, 95; tariffs on ethanol, 98, 101
U.S. *See* United States
U.S. Agency for International Development (USAID), 74
U.S. Geological Survey (USGS), 13, 31, 37, 80, 92–93

USGS. *See* U.S. Geological Survey
U.S. News & World Report, 21
USSR (Union of Soviet Socialist Republics). *See* Soviet Union

Varadero (Cuba), 123
Venezuela: costs of energy capabilities, 3; Cuba and, 4, 7, 37t, 45t, 51, 61, 77n10; economic issues of, 57, 81, 113; management of oil profits by, 117–18; oil exports of, 12, 13, 49, 55–56, 81, 117; national oil company of, 2; oil prices of, 12; oil strike of *2003* in, 6; productive capacity of, 121; United States and, 12, 111, 121, 122. *See also* Petróleos de Venezuela S.A.
Vietnam, 27, 35–36, 87

Wales, 75
Water utilities (Cuba), 76
West Germany, 50. *See also* Germany
Wind power, 53, 63, 64, 66, 113
World Bank, 105
World War II, 95

Zaarubezhneft (oil company; Russia), 27

BROOKINGS The Brookings Institution is a private nonprofit organization devoted to research, education, and publication on important issues of domestic and foreign policy. Its principal purpose is to bring the highest quality independent research and analysis to bear on current and emerging policy problems. The Institution was founded on December 8, 1927, to merge the activities of the Institute for Government Research, founded in 1916, the Institute of Economics, founded in 1922, and the Robert Brookings Graduate School of Economics and Government, founded in 1924. Interpretations or conclusions in Brookings publications should be understood to be solely those of the authors.